Learned in the Trenches

Maria Angela Capello · Hosnia S. Hashim

Learned in the Trenches

Insights into Leadership and Resilience
Compiled by
Two Women Leaders in Energy

Foreword by Hani Abdulaziz Hussein

Maria Angela Capello
Kuwait Oil Company
Ahmadi
Kuwait

Hosnia S. Hashim
PIC—Petrochemical Industries Company
Ahmadi
Kuwait

ISBN 978-3-319-62821-9   ISBN 978-3-319-62822-6 (eBook)
https://doi.org/10.1007/978-3-319-62822-6

Library of Congress Control Number: 2017947028

© Springer International Publishing AG 2018
This work is subject to copyright. All rights are reserved by the Publisher, whether the whole or part of the material is concerned, specifically the rights of translation, reprinting, reuse of illustrations, recitation, broadcasting, reproduction on microfilms or in any other physical way, and transmission or information storage and retrieval, electronic adaptation, computer software, or by similar or dissimilar methodology now known or hereafter developed.
The use of general descriptive names, registered names, trademarks, service marks, etc. in this publication does not imply, even in the absence of a specific statement, that such names are exempt from the relevant protective laws and regulations and therefore free for general use.
The publisher, the authors and the editors are safe to assume that the advice and information in this book are believed to be true and accurate at the date of publication. Neither the publisher nor the authors or the editors give a warranty, express or implied, with respect to the material contained herein or for any errors or omissions that may have been made. The publisher remains neutral with regard to jurisdictional claims in published maps and institutional affiliations.

Cover illustration: The picture on the cover is by Dr. Alexis Vizcaino, and portrays a detail of a fence of Palau Güell in Barcelona, Catalonia, Spain, a prime example of the exceptional creative contributions of Antoni Gaudí to the development of architecture and building technology in the late 19th and early 20th centuries. The beauty of this fence is also its strength: the interweaved irons. For us, it perfectly symbolizes the resilience needed by leaders in pursue of success.

Printed on acid-free paper

This Springer imprint is published by Springer Nature
The registered company is Springer International Publishing AG
The registered company address is: Gewerbestrasse 11, 6330 Cham, Switzerland

Foreword

No leader is so lucky or perfect, as to succeed without having to wade through myriad difficulties, usually unknown to others. I am impressed by the compilation of stories about visionary focus, grit, and determination that Maria Angela Capello and Hosnia S. Hashim have collected. Their perspective from interviews with these fine role models, who narrate work and life journeys in their own voices, is not only meaningful, but compels a deep reflection in each one of us.

Questions such as "*What would I have done in a similar circumstance?*" or "*Would I have had the courage?*" are a constant undercurrent in one's mind while reading every chapter. The whole dissertation provides a rich insight and an improved self-awareness for the reader, grounded on the life stories in this anthology. The authors invite us to share in their remarkable experience in interviewing these individuals, who represent the epitome of leadership and resilience.

I had the pleasure of working with Hosnia at different times in her career and was able to witness with pride, her progress, and achievements to the highest ranks of Kuwait oil and gas sector. I am similarly impressed by Maria and the transformational management strategies that she has used to launch innovative implementations in Latin America, USA, and the Middle East.

The introductory chapters are a brief preparation to what is undoubtedly the strength of this book: captivating narratives that allow the reader to experience the journeys toward success of these notable leaders. Each story is told in the first person, in the leader's own words. There are no filters that would hamper the energy and intensity of the stories. The reader is actively engaged, leading to increased comprehension of what qualities make a great leader, as the singular stories unfold with unexpected details, in some instances including life-changing moments.

Several of the episodes included in this book resonated with me, because I shared the experiences related during several of the events herein included, in particular, those anecdotes related to the oil sector of my beloved country, Kuwait.

The personal perspectives included by the authors in this book reveal a glimpse of the rich journeys of Hosnia Hashim, a pioneering leader and one of the finest role models in the Middle East, and Maria Angela Capello, a trailblazer in her own right.

Their book was envisioned and written in the Middle East, but it has global implications. It will serve to enhance the mutual understanding of the essence of resilience and leadership across cultures, across countries' borders, and across industries.

Kuwait City, Kuwait
March 2017

Hani Abdulaziz Hussein
Former Minister of Oil of the State of Kuwait
and former CEO of the Kuwait
Petroleum Corporation

Preamble

We met in Kuwait in 2008, when we both were working for Kuwait Oil Company. From the initial moment to date, we have established a collaboration that resulted in many pioneering initiatives in the oil sector of Kuwait that make us very proud. For several years, we have wanted to craft a compilation of the learnings and achievements attained in the numerous initiatives in which we were engaged together. We felt that the extraordinary match that we developed in our collaboration through the years was unique.

At first, we thought that such a compilation would provide an interesting reference for others. Perhaps, by identifying what worked for us and what did not, others would learn how to replicate this kind of winning approach such as the one that we built together. We then realized that we preferred, instead, to construct a more comprehensive learning platform for all those interested in leadership and resilience. With this book of stories about remarkable individuals, we want to trigger learning grounded on the reader's self-reflection about leadership and resilience. We want to share our conviction that reaching success is a gradual process, full of risks and decisions, and that it is as unique as each individual's story is.

This is how we formulated this book, as an assemblage of our own ideas on the corporate environment in relation to resilience and leadership, with profiles to demonstrate key concepts. We interviewed a small selection of our liaisons, colleagues, and friends, who represent a sample of a much larger network, many of whom have contributed over the years to our developing understanding of leadership and resilience. Their willingness to be interviewed for this initiative is as important as our own, because they helped to shape who we are today, with respect to resilience and leadership.

The leaders whom we invited to interviews come from different segments of the oil and gas sector, which is our business arena. We believe that the learnings and echoes derived from their stories are applicable to all industries, as leadership is truly universal. They were clearly "in the trenches," and we developed for them a higher respect than the one we already had, after discovering new details and hidden facets of their journeys toward their high ranks.

Being loyal to our original intention of capturing our fruitful learnings, there is a prolific utilization of the "*we*" and "*us*," and we definitively like that! Our choice of concepts, and our viewpoints and comments are the result of joint insights developed in our more than 68 combined years of experience in the oil industry.

We believe that a leader is like a lighthouse. A leader stands tall, guiding the ships into the shore with bright light, through quiet waters and during tempests as well. Inside each lighthouse, there is a steep interior flight of stairs, which has to be climbed one by one to get to the top to illuminate the shore. This book is about those hidden staircases.

We hope you will return to these pages many times, either to read again segments of this compilation for your own recollection or to share it with your mentees, as we encourage you to guide young professionals as part of your personal journey through life and work.

Ahmadi, Kuwait　　　　　　Maria Angela Capello and Hosnia S. Hashim
March 2017

Acknowledgements

We have learned a great deal from many people, and we express our limitless gratitude to all those who were key for our professional development, too many to mention. Separate acknowledgements recognize key individuals for each one of us, and we do hope all our liaisons will feel included, as it was our genuine intention to praise all of them.

We extend our special thanks and express our admiration to each of the leaders whom we interviewed. Each and every one provided a singular and unique tale of success. Our book was enriched by their stories and became a multifaceted compendium of learnings, narrated from different angles. Thanks to them, we completed our vision of leadership and resilience that we offer to the readers, with the hope it will trigger reflections and accelerate actions toward their own success.

Dr. Alexis Vizcaino believed in our idea from its incipient stage through fruition, and we appreciate his guidance during the publication process. We are very grateful to Susan Howes, who generously and skillfully edited our book, for her insightful recommendations.

Maria Angela Capello and Hosnia S. Hashim

Acknowledgements—Maria Angela Capello

Key individuals forged my resilience and leadership throughout the years. Dr. Mike Batzle boosted my self-confidence as a young professional. Dr. Martin Essenfeld, Jesús Patiño, and Dr. Giuseppe Giannetto grounded my managerial insights and instilled in me the needed resilience through critical professional growth phases during my years in Petróleos de Venezuela, PDVSA.

Badria Abdul Rahman and Abdulla Al-Sumaiti endorsed and promoted my participation as an executive consultant for the oil sector of Kuwait with a confidence in my abilities that honors me more than I can express. Their endorsement has been a privilege, and I cannot thank them enough.

My volunteer work for the Society of Petroleum Engineers (SPE) and the Society of Exploration Geophysicists (SEG) has been an enjoyable self-improvement and leadership track, where I encountered my fantastic role models Dr. Eve Sprunt, Susan Howes, Nancy House, and Anna Shaughnessy.

A special mention of heartfelt gratitude goes to Maria Jose De Sousa, who became my indispensable friend.

My relationship with Hosnia Hashim fills a definite chapter in my life! More than a boss, a colleague, or a friend, she has been pivotal in my own growth in the oil industry, and I am absolutely privileged to have partnered with her for this book endeavor.

The solid rock where I stand is my family: My sisters, Paola and Patrizia, have accompanied my paths in many essential ways. My dear mother, Mariesa, is an unparalleled source of maternal love and the daily strength from where I get my stamina.

My beloved husband and daughters, Herminio, Alessandra and Claudia, are my own perfect circle of love, and I have no words to describe their importance in my life.

Thanks!

Maria Angela Capello

Acknowledgements—Hosnia S. Hashim

Giving thanks to all relevant people in a career of many years in the oil sector is not easy, as to pick any name above the others may risk leaving out someone important. But with the courage to overcome this challenge, I want to acknowledge and appreciate all the individuals who shaped my resilience and endurance in difficult and complex moments, and they are in my list of mentors and role models.

A much reduced selection of my cherished liaisons through the years was interviewed for our book. I want to express my thanks to all and each one of them. Their words are the soul of this compilation.

My special appreciation goes to a wonderful person and devoted advisor who has been a catalyst for many of my achievements, Maria A. Capello. She has been for me more than a colleague an absolute and enthusiastic ally, who definitively became a dear friend.

I find solace in the memory of those who supported me the most, my beloved parents, who groomed me with endless love, and who shaped who I am today. Dearest thanks to my sisters and brothers, they are the solid foundation of my family and have supported me every single day.

Finally, my most expressive thanks go to my beloved husband and to my two beloved sons, Nawaf and Talal. They have propelled and keep propelling both my small and my big successes, supporting me with an unconditional encouragement and a decisive faith in my capacity that motivates me to be the very best I can be.

Hosnia S. Hashim

Contents

Foreword	v
Preamble	vii
Acknowledgements	ix
Resilience and Leadership Considerations from the Oil and Gas Corporate Environment	xvii
Before the Interviews	xxxi
Nader H. Sultan	1
A Glimpse	1
A Personal Snapshot	3
Arranging the Interview	3
A Shared Selfie	13
Post-scriptum	14
Hashem S. Hashim	15
A Glimpse	15
A Personal Snapshot	17
A Shared Selfie	31
Sami Fahed Al-Rushaid	33
A Glimpse	33
A Personal Snapshot	35
Arranging the Interview	35
A Shared Selfie	47
Post-scriptum	47

Maha Mulla Hussain 49
A Glimpse 49
A Personal Snapshot 50
Arranging the Interview 51
A Resilient Leader 52
A Shared Selfie 62

Dr. Kamel Ben Naceur 63
A Glimpse 63
A Personal Snapshot 65
Arranging the Interview 65
A Shared Selfie 75
Post-scriptum 76

Sheikh Faisal bin Fahad Al-Thani 79
A Glimpse 79
A Personal Snapshot 81
Arranging the Interview 81
A Shared Selfie 89

Dr. Ramona M. Graves 91
A Glimpse 91
A Personal Snapshot 93
Arranging the Interview 93
A Tornado from Nebraska 94
A Shared Selfie 99
Post-scriptum 100

Ali Rashid Al-Jarwan 101
A Glimpse 101
A Personal Snapshot 103
The Driving Force in ADMA 103
A Shared Selfie 112
Post-scriptum 113

Olivier Soupa 115
A Glimpse 115
A Personal Snapshot 116
Arranging the Interview 117
A Shared Selfie 128
Post-scriptum 128

Intisaar Al-Kindy 131
A Glimpse. 132
A Personal Snapshot 132
Arranging the Interview. 133
The Junior Geologist. 133
A Shared Selfie 141
Post-scriptum 142

Dr. Behrooz Fattahi. 143
A Glimpse. 143
A Personal Snapshot 145
Arranging the Interview. 145
From Iran to Iowa 146
A Shared Selfie 153
Post-scriptum 154

Fareed Abdulla. 157
A Glimpse. 157
Determination in Action 158
A Shared Selfie 168
Post-scriptum 169

Dr. Giuseppe Giannetto Pace 171
A Glimpse. 171
A Personal Snapshot 172
Arranging the Interview. 173
A Shared Selfie 185
Post-scriptum 185

Dr. Pinar Oya Yilmaz 187
A Glimpse. 187
A Personal Snapshot 188
Arranging the Interview. 189
A Shared Selfie 195
Post-scriptum 196

Lionel Levha 199
A Glimpse. 199
A Personal Snapshot 200
Arranging the Interview. 201
A Shared Selfie 211
Post-scriptum 212

Hinda Gharbi 215
A Glimpse. 215
A Personal Snapshot 216
Arranging the Interview. 217
A Shared Selfie 229

David T. Donohue, PhD, JD 231
A Glimpse. 231
A Personal Snapshot 232
Arranging the Interview. 233
A Shared Selfie 243

Dr. Nansen G. Saleri 245
A Glimpse. 245
A Personal Snapshot 247
Arranging the Interview. 247
Everything Is a Continuum 247
A Shared Selfie 258
Post-scriptum 259

After the Interviews: By Way of Epilogue 261

About the Authors 265

Resilience and Leadership Considerations from the Oil and Gas Corporate Environment

Working in any industry has the intrinsic challenges, unique to the sector in which it evolves, and it is a commonly accepted premise that if you like your work, you will succeed easier, despite the challenges.

But a detailed analysis will show that working hard is never enough and that leadership and success require great deal of resilience, accompanied by a privilege insight to capture opportunity, to be able to grow in the ranks.

In our compilation, the concepts about leadership and resilience will come from our own perception of these two traits, along the learnings expressed by 18 leaders we interviewed, that shared with us their own visions and learnings on the subject. Our experience in leadership and resilience comes from a combined 68 years of experience in the oil and gas industry, an experience that can surely be extrapolated to other sectors.

Of all industries, the oil and gas has characteristics that extend bridges to many other sectors, and these are as follows: the multinational nature of the business, the highly specialized profile of the workforce, and the cycles derived from the fluctuations of the price of oil.

The Multinational Nature of the Business

Reserves of oil are found in almost every country. Nowadays, with almost all countries producing oil and gas in large or small volumes, the accumulations of oil have evolved to become another concept. The resources of oil and gas can be considered a commodity that dependent on the level price could or not be subjected to production.

This situation creates a truly multinational outreach for the technological solutions, as all oil fields eventually need water, CO_2 or steam injection, artificial lift, and other recovery mechanisms and production enhancement processes. So, technology providers, oil field service companies, and contractors have already worked around the globe and established offices, workshops, and training centers globally.

Likewise, the exploration and production companies, in the old times mainly operating with a focus on a reduced number of countries with huge reserves of oil, faced challenges in the seventies, to embrace and accept as part of the new business scenario the incumbent NOCs, National Oil Companies, and the OPEC, the Organization of Petroleum Exporters Countries. This new scenario at the time revolutionized completely the scheme of things, and the balance of the business in the oil sector changed forever. More recently, after a major crisis, in 2008, the industry embraces a new change that started to evolve rapidly from 2010 and onward: the production from shales, an unthinkable event that again reshuffled the markets and positioned the USA as a player in the export market, presenting a new challenge to the usual providers of crude in the planet.

The corporate environment follows this dynamic, and currently, every company, public or national, producer or of services, consulting firm or private, dedicated to the oil sector, has a multinational workforce and generally operates in several countries and continents.

Leading a multinational workforce presents challenges that are not always visible to an untrained eye. The leaders need to handle communication and motivation in different ways with different sets of people, dependent on their culture. The very essence of leadership is challenged by multiculturalism because leadership is not only vision, experience, decision capacity, or technical knowledge. It is very much communication capacity, which needs to be changed dependent on who will receive the message.

Our learnings in the oil industry, handling, influencing, or leading a multinational workforce in a variety of countries, point to highlight the need of tolerance and respect as fundamental traits to success. The leadership is perceived differently in different regions. And if leaders are highly regarded and frequently almost revered in regions such as the Middle East and Latin America, in other areas they are severely scrutinized and are the target of sour criticism. The leaders then adapt and need to resist to succumb to conceit in one extreme, or doubting their self-worth, in the other.

Additionally, they need to become translators and communicate adequately with different cultures they deal with, making sure the business objectives are met. In meetings with few individuals or in presentations for town-hall large audiences.

Multiculturalism triggers also a special kind of resilience: to keep attempting better ways of communicating with different cultures and understand the hidden signs of acceptance or rejection from people of different cultures and to navigate "by instruments," in many occasions that may result in an unfamiliar situation.

The Highly Specialized Profile of the Workforce

Many industrial sectors have a high demand and utilization of technology, and let us think for example the aeronautical sector or the telecommunications. The oil sector has traditionally been an industrial area of massive utilization of technological solutions.

It is a fact that the oil sector is a slow adopter, but under any scrutiny, it is also a massive consumer of technological solutions, and once any innovation is proven to impact positively oil production, it will be extensively applied on a global scale. When the fresh graduates enter the oil industry workforce as new hires, they are submitted to extensive training programs, of 3 to even 5 years of duration before they become fully operational. This is a reality, since the academia prepares the engineers and other scientists employed by the oil sector with a basic and conceptual knowledge. But the instrumentation of that basic or conceptual knowledge into practical implementations at work requires a long period of complex processes and schemes of training, self-learning, mentoring, and on-the-job training, so that the employees become fully operational.

Furthermore, the near-term future will require even higher specialized profiles, as incrementally, the oil fields demand more technological solutions to sustain production, requiring tertiary production schemes and sophisticated subsurface studies and drilling operations to drain the reservoirs with a smart approach, not damaging what is left.

These requirements set many challenges for the technical management and human resources or training departments of the oil and gas companies, as incrementally, the search for the adequate profiles becomes more and more challenging, looking for professionals with the best preparation, multinational background, and willingness to relocate.

The capacity of relearn has become more important than the capacity of learning. Professionals able to leave behind old ways of work, to embrace new technologies and their pertinent workflows, will advance more rapidly than their colleagues who stick to their way of work or technological solutions. This applies to all industries, as technology becomes more and more an integral part of the workflows in every sector.

Back in 1997, Maria was invited to Statoil for their internal Annual Symposium as a Key note speaker and to the "Recent Advanced and Road Ahead" session of the SEG, Society of Exploration Geophysicist's Annual Meeting. It was to present what was she leading in PDVSA Research Center for 4D or time-lapse seismic. In a time when repeatability issues were starting to be a problem to characterize the response of 4D analysis, Maria's research team was already planting permanent geophones in Lagunillas field, near Maracaibo Lake in Venezuela, and monitoring steam injection impact in seismic attributes. She left behind what was usual in seismic acquisition to test new ways of enabling repeatability of the seismic 3D acquisition process for time-lapse purposes. She was developing her own leadership profile, but most interestingly, the leaders of her organization provided the necessary endorsement, confidence, and support, so that she could pave the way toward new advances in her technical field of interest. This is the kind of leadership that is nutritive of a highly specialized workforce. A degree of freedom, based on trust in the technical and leadership capabilities of the individuals deserving it, is needed to trigger yet new levels of leadership. Other experiences later on sustained this vision that managerial support and freedom is indispensable to foster leadership.

Another example of the highly specialized profile of the workforce and the challenges it poses to leadership comes from the shift that Hosnia applied in the drilling strategy in the North Kuwait Directorate of Kuwait Oil Company, when she was the deputy Chief Executive Officer of that asset. In 2010, all the wells in the drilling plan were vertical, and she demanded to restrategize the drilling plan, to consider more efficient (and modern) ways of drilling, to maximize production, and to optimize the use of surface land. It was a challenge, as the workforce was not used to the new schemes of multilateral wells, horizontal wells, and advanced completions in several intervals for each well. The shift to the new strategy required parallel actions taken in training, awareness campaigns, and the hiring of specialized companies to support KOC's own efforts in this shift toward new drilling schemes. Clearly, the workforce has to be uplifted in parallel with the technological changes accompanying every industrial sector, and leaders need to be aware of the technological trends, so that they steer the efforts to reach new technological levels and be able to match the competitors. Effective leadership enabled these changes.

The Cycles Derived from the Fluctuations of the Price of Oil

The price of oil fluctuates. And not by small variations. It fluctuates rapidly and harshly, seriously impacting the business. It is no secret that the oil and gas industry is hit by the reactive fashion with which we respond to the fluctuations of the price of oil. In consequence, the profile of experience and age of the workforce experiences variations as well, inserting gaps not always beneficial to meet the demands of the business or the dynamics of the sustainability of oil production.

We find our organizations lacking the technical capacity, after massive layoffs are actioned, especially by service companies and international oil operators, seeking to reduce costs to benefit their stakeholder's profits. But this approach to business has left our industry helpless to cope with the demand of oil when and if the price returns back to profitable margins. It is a story already experienced, but not learned at all. In 2016, the industry again let go valuable minds and hearts of people committed to the industry. In 2017, when the price returned to profitable levels, we witness recruitment campaigns at a global scale.

The cyclic nature of the industry of oil and gas is shared with other industries, like the touristic and academic sectors, which follow the major economic trends, generally tied to the oil and gas price and movements.

When companies and organizations do not have stability in their intake and attrition numbers, but behave in response to the waves in the market, dependent on prices, the management and leadership experience additional challenges. Challenges that derive from having to produce the same or more volumes of products with less personnel, many times, without key personnel. Also, they have to deal with shifts in the driving forces that command production and marketing, changed by the new dynamics of the low oil price. The big gaps in expertise that these low-price periods produce in the oil industry are difficult to fulfill, and their effects remain for decades. The industry definitively owes its resilient leadership to several of these difficult low prices of oil cycles.

Lessons learned from the cyclic nature of the oil industry resonate loudly in other sectors that have kept their key expertise live and within their ranks.

The resilience of many oil industry leaders comes from having survived successfully several of these cycles, reinventing themselves. These leaders faced serious and many times extreme challenges in their organizations discovering within themselves the strength and courage to launch entrepreneurial initiatives, alone or with partners, in what is a very difficult path that propelled the best in them to flourish.

Leadership and Resilience

The main purpose of this book is to trigger a reflection in the reader about what are the qualities that may be associated with leadership and resilience, and in this chapter, we will provide our own vision about these two central and pivotal elements for a success journey in any profession.

The reality common to so many leaders is that each person is unique, and unique is his or her trail toward success. So, to try to define a leader with a list of characteristics will be to try to put an infinite variety of styles into a single mold, and that is certainly not easy, nor feasible.

Nevertheless, there are certain qualities that transcend the individual realm of the uniqueness, to shape platforms that are associated with success and leadership. And with resilience, we will expand those qualities, with the hope that the understanding of their importance would enable a determination to pursue an ownership of those, so that they become a daily commitment, until they are imbedded in the character and personality of those interested in pursuing accomplishments in life and work.

Resilience

Resilience is the process with which we face, adapt, and overcome adversity and defeat. Recover from difficult experiences is to be resilient.

It would seem an extraordinary quality, but it is available in every single person. When trauma or stressful situations occur, the person submitted to those experiences distress and, in many cases, also emotional pain and sadness. The process to overcome those feelings and identify the coping mechanisms and winning strategies to win over the struggles and succeed in enduring is resilience.

Resilience is available in each one of us, and we can learn to become resilient. How? We become resilient when we develop behaviors aimed to recognize our own value and to realistically ponder the situation, so that solutions become visible.

There are several factors important and recognizable in the process of becoming a resilient person and a resilient professional. We would like to highlight those that we recognize have added value in our professional life or that of our role models:

External factors important to develop resilience:

- **Support system**, with relationships that are based on trust, at work and within the inner family circle
- **Role models** that foster self-confidence, at work and in life
- **Availability of sponsors** (champions), who are sincere promoters and experienced in identifying and promoting the intrinsic value in the individuals they sponsor
- **Availability of assertive, fair, and transparent feedback** on shortcomings or failures related to recognition processes such as promotions, awards, or incentives based on merit, to enable open communication that helps in the identification of enhancement paths

We add value to our life and our professional career when we surround ourselves with factors that will enhance our resilience. Questions such as "*Do I have a sincere, loving support that I trust?*" or "*Who are my role models?*", among others, are questions that we need to constantly ask ourselves in complete sincerity. And we need to seek remedial actions if we find negative or inconsistent answers. It should be our priority to ensure we have a strong support system around us, because that peace of mind will make us stronger, more resilient, and at the end, winning individuals. And we have to opt for systems that provide us with a genuine feedback, one that is not sugar-coated to please us, but that assess the weak and strong aspects, enhancing our own perception of ourselves. It is only with honest feedback that we learn where and how we failed and where and how we succeed. And many times, that feedback can also address the most difficult and intangible question of why we failed or succeed.

We need to face our environment with a critical eye, realizing the importance that fairness and honest feedback have, so that valuing those, we maintain an alert and vigilant attitude, seeking that kind of feedback constantly, to boost our resilience in adversity cycles.

The commitment to develop our resilience is a very personal choice. Because resilience can be developed, and besides the external factors discussed, perhaps the most important aspects needed in forging resilience are the intrinsic or internal factors.

Other external factors of relevance to promote resilience are the selection of our role models and sponsors. A role model that motivates in us better attitudes at work, and models for us a balanced perspective at work, is always the fastest way to better ourselves, as there is nothing wrong in imitating winning models.

Role models are elected by us. So, pick high and pick the best role model you can imagine from a variety of role models that are nearby and accessible. No one precludes you to pick any role model of the highest caliber there is. Study him or her and analyze what makes this person successful and what you can imitate and implement in your life that would reassemble that person, that professional you want to become. And from your accessible role models, seek advice, seek guidance, and ask the whys and the hows. They will help you, and most probably, you will notice that most importantly, they also had to persevere and be resilient to reach to the level they have reached.

The sponsor is a person who takes on her the commitment to promote your name and your role, even if you are not present. She champions you in meetings where promotions are discussed or opportunities envisioned. A sponsor is the best thing that can happen in the professional ladder, as many people have mentors and coaches, to take care of their long-term professional life, or technical skills development. But very few have individuals that will fight for them, endorse them, and advocate for them. Recognize your sponsors. They may be out there, and you have yet to identify who they are. Thank them, and ask advice to them to become a better professional, a better person. We have never seen a sponsor that would not be willing to walk the extra miles needed to support their championed individual. So, do not hesitate and ask how to be more resilient to that person who sponsors you.

Finally, the last external factor that we need to care about is to assess whether the organization we work for has in place assertive, fair, and transparent feedback processes. This is especially important when we face our shortcomings or failures, in low cycles of performance. We all go through good, optimal, and excellent cycles, but also through low moments, when our focus or our attention to detail and quality decays, for a variety of reasons, resulting in a less-than-optimal performance. A transparent process of promotions and recognition of performance should differentiate between the "*Who am I*" versus the "*How I did*," and explaining that to the individuals is critical, as not to harm self-esteem or jeopardize a healthy resilience, that enables the individuals to jump back on track and start over or enhance their performance in difficult situations. The loss of awards or incentives based on merit that we thought were deserved and did not arrive can be harmful, and there is nothing more painful than seeing others enjoy the benefits, recognition, or status that a loss recognition of our performance brings. It is then important to bounce back and exercise our resilience, because it is thanks to this quality that we endure in our pursuit of success.

If the organization where one works does not have a fair and transparent system of promotion and recognition, it will be difficult to grow a healthy resilience, one that compensates the lows and highs of the natural life and work cycles of an individual through time, regaining with so the confidence in the larger system.

In our careers, there have been occasions when we were not given the promotion or the salary rise we expected and we thought we deserved, but it is also truth that the resilience this grew in us enabled the force to try harder and better in the next loop, in the next post, achieving the aimed targets in the second opportunity. But external factors are not all what is needed, and in many opportunities, we cannot change the environment or system surrounding our lives and work. This is when the internal factors of resilience acquire a special relevance.

Internal factors key to develop resilience:

- **Positive attitude at life and work**
- **Self-knowledge of weaknesses and strengths**
- **Emotional intelligence**
- **Assertiveness**
- **Self-control**

Developing resilience from within is a personal journey. People do not react in the same way to traumatic and stressful life events, or to minor adversity events.

An approach to building resilience that works for one person might not work for another. People use several strategies, and the factors we listed above are what we consider to be the most important in terms of developing that peculiar strength that accompanies true success. We have seen it in the leaders we interviewed in this compilation, as well as in other role models we have met in our careers.

Firstly, to have a positive attitude at life and work really helps and triggers the endurance and strength needed to keep on, when facing adversity. It is very easy to criticize, but very difficult to build, to innovate, to see solutions and opportunities where others only see problems. And this cannot happen if we maintain a negative attitude.

We are not referring to an unfounded optimism, of foul, shallow people, but to the positive attitude of the dreamer, of the achiever, that sees the goal in front of him or her and, at the same time, feels the strength within to fight any battle to accomplish the set goals. This is the kind of positive attitude we are referring to. This is a quality that naturally builds and enhances resilience.

We are certain of it, because we have seen positive thought in action in our own careers and that of the people we admire the most.

One can learn to develop a perspective at life and work that is intrinsically positive. And if that does not come naturally to us, we can force it on ourselves, until it becomes a habit. Some of the interviewed people did just that by asking themselves "*what is the worst that can happen?*" and realizing their own luck, strength, and opportunities ahead, radically changing their point of view and initiating a new path at work that led to success.

Another element we listed is to develop our knowledge of who we are. Deep inside us, there is our true self, and we have to know our strong characteristics as well as our weak ones, with honesty and clarity. No one knows oneself better than himself or herself. We have faced alone those difficult moments, and we know how much we can sustain a crisis, many times surprising ourselves with an endurance we did not know we could apply.

That is because we grow during each crisis. As difficult as it can be, the critical moments in life and at work teach us who we really are and how strong we can be. A death of a loved one, an accident, an unexpected external crisis, demands from us the best of our fibers, and it is in those moments when we develop our resilience. Even if always painful, we need welcome our crisis, as we gain from them.

Some words are needed to discuss our emotional intelligence, assertiveness, and self-control. The adult behavior is not as common as one may think, and generally, we are surrounded by professionals and people who did not grow beyond the childhood or teenage years in their attitude, perspective or take at life and work liaisons. It is then very difficult to handle situations where the assertiveness is not present, but instead, we have to deal with a continuum of demands and expectations not grounded on reality.

The practice of an adult attitude, with the proper confidence, or assertiveness, elevates the discussion to a rational plane, leaving behind impulsive or precipitate behaviors, favoring resilience as a natural consequence.

Resilience is in one word, the capacity to endure and bounce back in adversity. And that can be exercised. We do not wish for you a path filled with obstacles, but we recognize that the obstacles we have faced are what made us strong. May you be the strongest and most resilient person you can be.

Leadership

One of the most popular topics in business scholar articles and in self-empowerment books is leadership. We do have an intuition of what leadership is and what it is not. For example, when we see a recently appointed manager give orders or impose activities, we identify a person that will not succeed as a leader. But then, what are the traits that trigger so many articles and books? Are there common qualities or traits at all that define a leader? Could those be taught or are leaders born with a predisposition to be leaders?

There are no simple answers to these questions. So, we will provide our own vision on leadership, and we would like to summarize that for us, leadership is the compendium of attributes that enable to lead people, and not things or activities.

It is all about leading people.

In our perspective and experience, remarkable leaders are those capable to inspire people and see the best in each individual to energize the whole team toward common goals and, most importantly, exceed the individual expectations to collectively achieve those goals.

Managers deal with actions such as monitoring, implementing, planning, ruling, checking, commanding, directing, whereas leaders envision, inspire, create, listen, solve conflicts, partner, create bonds, and most importantly, coordinate, enhance, and align wills toward specific goals.

If we were to list the qualities that identify leadership, we would fall into a trap of systematizing the approach to leadership, and we consider each leader is as unique as their individual trail of success. But there are certainly common platforms from where the remarkable leaders grow and found themselves as lighthouses, signaling the path, and establishing the north toward where to navigate. They are capable to do this for themselves, but most importantly for the teams that they lead.

It is not about titles or designations. Leaders may be found in assigned leadership roles, but also in those key individuals of an organization that do not possess formal power titles, but around whom the teams centered and gain energy to move upward and onward. Many of the leaders we include in this compilation discovered their leadership early in their lives, in activities not related to their future career, as summer jobs, volunteering jobs, or sports. Others developed their leadership through the years, being exposed to leading opportunities in critical situations.

The qualities we chose to list and comment here are not all the ones showcased by the leaders we know so well, nor all of those leaders that we interviewed. It is our own summary.

We decided to list and provide a few comments on what we consider to be foundational qualities of leadership, integral to shape a leadership platform. From this platform, some will be able to rise from the average leaders' cohort, to become remarkable leaders. We hope that the reading of the individual trails of success of each leader that we included in our book will activate your own conclusions about what are the qualities of a truly remarkable leader, and what is the combination of opportunities and qualities that shapes a brilliant career.

Vision
The capacity of seeing and imagining a better, improved, or enhanced future for the organization, team, or group is one the main qualities of leadership. The clarity with which a leader sees where to go is the main difference with other individuals, who do not have the ability to envision as a target a situation different than the current one. To envision a future is not enough, and the true, remarkable leaders also envision how to get there and with what resources. The vision of leaders also encompasses their capacity to identify alignment paths and to optimize the resources and time frames for the tasks at stake.

Integrity
Leading others cannot be sustainable without being moral, ethical, and honest, the pillars of integrity. Integrity is that word that encompasses the worthiness and correctness of an individual, and more so of a leader. Nothing is more catastrophic for a leader than to lose her or his integrity, as it is the cornerstone of leadership. A value and quality learned from family values and as part of the very early-years education, it is imbedded in the fiber of leaders, and we would say it is immutable through the years.

Commitment and Accountability
To be accountable for our own actions is a given part of our professionalism. But leaders go a step further and feel and are accountable for the actions of every individual in their organizations. They take pride in being responsible, and monitor the performance but more so the workflows and tools available for their team members, to provide them with the needful to achieve the goals, overcoming the challenges on the way. Commitment to the organization's goals is an integral part of their leadership, and many times they set those goals, as they envision the future and shape the present to reach there.

Optimism
We have never met a leader that is not optimist about the future. They must be! Because they are the enhancers of the present to reach the vision they have, that perfects the path to achieve the goals and targets of the future. The optimism they reflect is a shiny one, one that creates waves of replication and resonates within the team, elevating their team vision to become also optimistic and hence instilling positivism and invigorating their contributions.

Self-Confidence

The remarkable leaders have a self-confidence that characterizes them beyond other traits. You are with them, and you feel more confident as a consequence of looking for their advice and support. Their confidence is contagious. Their confidence is also genuine, as it is based on results and achievements that they have piled up, and it is a result of their convincement they can achieve the vision that many times, only them have established and understood with total clarity.

Communication

It would be absolutely pointless to have a vision, to be committed to it, to have accountability, to be confident about being able to achieve it, if a leader cannot communicate with his or her team. The communication ability of a leader is one of the most remarkable traits that propel them to success. Communication in every sense:

- *Lateral*: To liaise with his peers and colleagues within and outside the organization. The capacity to be "one of the pack" is crucial, as leaders easily may get adversaries among peers who compete for top roles, and competition becomes harder at every step upward. Remarkable leaders collaborate and gain from the collaboration among peers, understanding and leveraging on the value of this approach.
- *Cascading*: To communicate the vision, objectives, challenges, and path to the team members. This many times is a capacity to literally translate those to all in the team. An operator needs to hear from a leader the corporate goals in a simple language, applicable to his or her daily tasks. A laboratory technician requires to understand how the corporate goals are applicable to her daily routine, and here, the translating capacity of the leader boosts his or her efficiency. The remarkable leaders are also excellent storytellers and develop an intriguing capacity to adapt the core of their key messages to all sorts of audiences. They feel confident communicating in reduced circles as well as in large audiences.
- *Upwards*: Liaising with supervisors when you feel inside of you the strength of your own leadership is not always easy. This is why the communication of leaders with other leaders who are steps above them in the leadership ladder is significantly important. Humbleness and self-consciousness become increasingly relevant in this segment of the communicational ability of leaders. It is in their interest, and they do it magnificently, to communicate their vision, their goals, and their expectations with clarity. But it is more significant the ability they have to communicate their needs, and those of their teams to achieve the set goals. In the communication

with their bosses, the leaders also ask critical questions, without filters, and provide their own insights on challenges and strategic decisions, many times becoming indispensable for the organization, that start to count on them for taking important steps into the future. It is precisely this capacity that many times triggers their own growth in the organization.

Before the Interviews

Our experience and insights have been molded and augmented by hundreds or thousands of interactions with people at work and in life. Some are already in the past, in other countries and roles, many years ago. There are others, with whom we interfaced more recently, in executive roles in leadership and consulting in the Middle East oil industry. We are shaped by a life of globetrotting, in work and volunteer activities related to the oil industry.

We are very proud of the colleagues and liaisons who helped us to climb the stairs of self-improvement. Step-by-step, we climbed our own stairs. It was easy at times, but usually difficult, almost always. We have not yet arrived at the end of the stairs, but the path has been facilitated by the support, example, guidance, and advice of many. Too many to mention. Too many to count.

We are convinced that our individual experience would have not been enough to capture the essence of what is leadership and resilience. The collective experience of these extraordinary leaders throws a more brilliant light to matters pertinent to resilience and leadership, as each leader's trail of success is unique.

This is how we envisioned the idea of inviting some of our cherished colleagues—many of them have become already dear friends—to help us in this compilation about leadership. They generously provided not only a swift positive response to our request, but also dedicated their time and attention to the interviews. They shared with us their own vision, stories, and even anecdotes, amplifying our storyline with new sounds and volume. It is the inclusion of their respective visions which makes this book a rich, multi-faceted composition with several layers of experience.

Their insights were provided as a result of informal conversations, shaped as interviews, following a common path created by a similar set of questions. The result was an incredible variety of perspectives, as unique as each one of our interviewed leaders are, providing variations on the themes of leadership and resilience.

We thought all interviews would result in common stories. We had developed a preconceived notion that the careers of these outstanding leaders were much more facilitated than our own careers had been. We assumed that they received strong support from higher leaders and that they were provided with a continuum of opportunities which were granted easily. We had this perception because we were—as everyone else—in great admiration of the shine of their great success. We did not know the details of their individual struggles and the intricate net of decisions and circumstances each of them faced in their careers. We were wrong, and we discovered each one faced difficulties that many times seemed unsurmountable, but which also strengthened their resilience and spirit to reach higher accomplishments.

Some of our interviewed leaders shared their inner self, when tapping into memories of the past, aiming to explain how they reached their success, how they made decisions in important crossroads, or what they left behind as trade-offs. Tears were shed, and laughs were shared, in what were often intense moments of sharing and learning.

Nader H. Sultan

"I was able to develop the ability to stay calm during a crisis".

A Glimpse

There are crises very similar to storms, where an anchor is a salvation. The State of Kuwait has navigated through several storms. The top one without any doubt was the invasion the country suffered on August 2nd, 1990. The country was hit in many ways, with its oil industry, in particular, being affected in unconceivable ways. In those dark days, from the regional headquarters of KPI in London, Nader Sultan became one of the valuable Kuwaiti anchors. Among the leaders of the wonderful collective effort, Nader Sultan was a key strategist, who was crucial in the efforts to manage both the Kuwait oil assets abroad in its worst moments, and assist in the planning for the return to Kuwait.

With particular strategic insight, Nader's participation in the oil and gas sector, not only in Kuwait, but on a global scale, continues to shape a remarkable career as a global leader in energy, leaving his mark indelibly imprinted in the Middle East oil and gas industry.

He occupied the highest rank of the national oil company of Kuwait, Kuwait Petroleum Corporation (KPC), as Chief Executive Officer in 1998, an appointment that came as a result of the many accomplishments accumulated earlier. He was Deputy Chairman and Managing Director of Planning in KPC; Executive Assistant Managing Director for worldwide product sales; and President of Kuwait Petroleum International Limited (KPI), coordinating the international refining and marketing operations of KPC.

In 2004, he retired from KPC and became involved in a number of initiatives for the advancement of the new cohorts of leaders of the Middle East. He is the Director of the Oxford Energy Seminar at St. Catherine's College Oxford University, which has become a prestigious hub for the networking and leadership growth of top leaders from important oil-producing countries.

Externally more active than many other executives, Nader Sultan is the Chairman of Ikarus Petroleum Industries, and serves on the Board of Governors of the Oxford Institute of Energy Studies, the International Advisory Board of Dana Gas, and the Board of Directors of Fluor Corporation.

- Graduated from the University of London with a B.Ss. in Economics.
- Joins KNPC as executive trainee in 1971, and grows in accountabilities, as KNPC Assistant Regional Sales Manager, Manager for International Marketing, Sales Manager for West of Suez region, and Supply Manager.
- In 1980 is appointed Executive Assistant Managing Director Worldwide Product Sales, KPC.
- 1983, President of Kuwait Petroleum International Limited (KPI), responsible for coordinating international refining and marketing operations of the KPC holding.
- 1993 Deputy Chairman and Managing Director of Planning of the Kuwait Petroleum Corporation (KPC).

- 1998 is appointed CEO, Chief Executive Officer of KPC.
- 2004, founded F&N Consultancy.
- He is the Chairman of Ikarus Petroleum Industries. Serves on the Board of Directors of Fluor Corporation, Dana Gas International Advisory Board. Previously, had served on the advisory board of Riverstone Holdings LLC, the Supervisory board of Al-Markaz Energy Fund, Advisory Board of Schlumberger Business Consulting and EMEA Advisory Board of Nomura International.

A Personal Snapshot

Nader Sultan is a well-known figure in Kuwait. There is no important energy event that would not invite his presence as Moderator, or Panelist. With plenty of reasons to receive his guidance, insight or recommendations, the current leaders in the Kuwait Oil Sector occasionally seek out Nader with confidence and unbreakable trust.

Hosnia knew about Nader Sultan early in her career, as Nader had already an important leading role when she started working in the oil industry. More specifically, when she started to progress into the executive roles of Kuwait Oil Company, with the Oxford Seminars as an integral part of the leadership training of the Director-level leaders.

On the other hand, Maria, had learned about Nader Sultan during her tenure in Kuwait, as she has been the Coordinator for the Society of Petroleum Engineers (SPE) Conferences, HOCE Heavy Oil Conference and Exhibition 2013 and 2015, and KOGS, the Kuwait Oil and Gas Show 2013, 2015 and 2017.

The style and simplicity of Nader Sultan makes of him an approachable leader. His story and trail of success speaks of resilience and leadership, in that order.

Arranging the Interview

It was a great joy to secure a conversation with him, and to include it in our book. We were not sure if he would have the time to commit, not only to the initial exploratory conversation, but also to the loop of reviews and possible editions needed for a final version of his chapter.

We scheduled two interviews to learn about his stories, anecdotes, and challenges. The time was not sufficient, and we wish to continue this exploratory path and know more about Nader Sultan in future opportunities.

Nader is an exceptional storyteller, and we discovered that no questions were really needed, because he unfolded the threads of his stories without further cues, in a logical sequence. We feel that the summary hereby presented is just but a glimpse of a greater story of growth, one that has not yet finished.

An Anchor in the Storm

Early Challenges

Tell us about the initial times in your career, Nader.

My mother and father temporarily left Kuwait in 1937 to reside in India, initially for my father to study for a Masters degree. There were only about ten Kuwaiti families in India at the time, and we lived there for quite a while.

We came back to Kuwait in 1958. At that time, my father was appointed Manager of Public relations in Kuwait Oil Company (KOC). Having perfect fluency in English, with a Master Degree, he was considered ideal for the post. So, he settled in Kuwait and for a short time, I went to school in Ahmadi, the compound town around the Burgan Field and Offices of KOC. My parents considered, following our education in India, that boarding school was a good option for the future education of my brothers, sisters and me. So, I was sent to boarding school in the UK, to Gloucester, and then I went in London for my university studies.

What did you study?

I studied Economics in the University of London, and graduated in 1971.

How did you choose Economics?

As any other student, I was not certain of what was my preferred career, and I liked so many subjects! What I wanted was to study a subject that would be general enough as to encompass many possible activities in my career, and Economics fit that purpose just perfectly.

Once you are in Kuwait, you realized everything is related to the oil industry.

Yes, but an Economics degree is not the most direct way to get involved in the oil industry, and I am curious. Although it might be redundant to ask this to a Kuwaiti person, given the importance of oil for your country, please tell me … How did you become involved in the oil industry?

From the very beginning of my career.

When I graduated, I met several Kuwaitis who were just arrived in London, to open up a new office for the Kuwait National Petroleum Company (KNPC), which handled the refinery newly built in Kuwait Shuaiba. At the time, KNPC was a partially private company. The London office had a specific mission, which was to handle the international marketing of the refined products from this new refinery.

The Dream Job

The people newly hired would be trained in Kuwait, for 3–4 months, to then be positioned in London. KNPC was looking for people, and possibly for Kuwaitis, with preparation in economics or marketing, for their international marketing department.

I applied. I was interviewed. I was hired!

You can imagine, a young professional, sent to work in London with a good salary, was the dream job. It was exciting. Travelling to Singapore, Indonesia, Australia and Bangladesh. Such a rich experience in every sense. The people you meet. I was selling refined products on a global basis. You would pay someone to do that job!

I had to take care of marketing and sales for KNPC firstly in the East, then later in the West, and this gave me an incredible exposure to different companies, cultures, and managerial styles. I acquired a network and managerial ability in a much reduced timeframe, logically imposed by the need of closing contracts, meeting our marketing targets and expanding business circles for the company.

Those were indeed remarkable years that started to build my own insight on the oil industry, especially from the international marketing point of view. Many years later, Kuwait was in deep trouble. But we will talk about that.

A Young Executive

In 1980 I was appointed Executive Assistant Managing Director, Western Hemisphere Product Sales of Kuwait Petroleum Corporation (KPC). I was only 31 years of age. I consider I was extremely lucky and blessed with this opportunity. It was a colossal opportunity!

Those were different times. I was appointed EAMD at a very early age, and this fact exposed me and my peers, like Hani Hussain who was also appointed at the same time as an EAMD, to managerial challenges and decision making paths that were truly difficult, which shaped our careers later on.

That early exposure to executive responsibilities and decisions shaped who I am today. It was foundational to any leadership trait I developed. It was a full-immersion, hand-on, practical approach to leadership.

In 1983, I was appointed President of Kuwait Petroleum International (KPI), and was responsible for coordinating the downstream operations of the KPC holdings in Europe. This was after we acquired the downstream assets from Gulf Oil. We were immediately in a completely different competitive field, and had to develop our strategy to compete against the long established big players in the downstream market.

The large majority of the downstream assets we bought were from Gulf Oil Company, so at the time of the handover, as President of KPI, I went to meet with the president of Gulf Oil in Europe. At the time, I was 36 years old, he was 55 years old, so I thought he was an old man!

He gave me several good recommendations. Most importantly, he told me that he had reviewed my profile and background. Realizing I came from international marketing, and that in this new role, with 6 countries under my responsibility, he saw that I would not have any more the time to directly meet with my customers, as they were now thousands of retail customers. His guidance was that instead, I would have to work through my managers and work through people to deliver the results I needed.

That was a competency I had to develop. And quickly!

One Funny Moment

Maria, at this point, I want to tell you about one funny moment in my career, because it is related to this meeting with the Gulf Oil President. I will always remember this anecdote! It was just before I met him, I went to the Gulf Oil building reception desk.

What happened?

At the moment of presenting my name to the receptionist/security man, he looked at me and said; "Oh, that is interesting we have two people called Sultan coming today". "Who is the other one?" I asked. "Oh! The other one is the President of Kuwait Petroleum International".

The receptionist was expecting an old man, a more "presidential" profile as per their view. Not a 36-year person such as myself.

I never corrected him.

A New Brand: Q8

Nader, I think this is a good moment to ask you about your major accomplishments. Tell me, what makes you proud when you look back at all what you have accomplished?

There are a couple of things of which I am extremely proud. It was a result of an extraordinary teamwork, which I was fortunate and privileged to lead, for the Kuwait oil sector.

The first achievement was the branding of our products in the international arena.

It started like this. We needed to ensure our competitiveness in the downstream sector. I mentioned earlier that we had acquired the downstream assets from Gulf Oil. So, Gulf Oil had granted us five years of use of their brand by contract, when we purchased the downstream assets. So, we launched a marketing study, using focus groups and other techniques. The results clearly showed that a new brand was needed. The Gulf brand was not well liked and we would have difficulty competing with the brands of Shell, Esso, and others.

We decided to create a new brand logo and name. It took us 10 months of detailed studies to decide on the logo. Because of Kuwait's heritage, we decided on a colorful version of two sails. Then we had to choose a brand name.

The marketing company then presented us with about 35 names from which to choose. I remember as if it was yesterday. They give us names like Dana (which means pearl, in Arabic), as well as so many others. Towards the end of the list, we glimpsed "Q8". It was a western way of saying Kuwait, but indirectly. We really liked it. Everyone we showed it to liked it! It was good in all languages, so the transfer across cultural and country barriers would not be a problem. We repeated "Q8….Q8…". Yes, that was it! We had picked our brand name.

The launching of Q8 was a very exciting phase of my tenure as President of KPI.

We knew we had a success when we presented the new brand to our dealers in the Benelux. They were very talkative but one man at the back was quiet throughout the presentation. He had been a dealer for Gulf for 30 years. When asked about his view, he said in French "*Absolument magnifique*"! He loved it!

There is another interesting aspect of what we achieved at KPI—This was an observation made by one of our competitors. He was surprised by how we had been able to put together and retain a senior management recruited from different European major oil companies.

It was rare for executives to leave these companies in the 80s. I think it was the exciting challenge they saw in our company. Owned by an NOC, resource rich, launching a new brand, trying to take on the big established competitors.

Seizing the Future

And what other achievement do you remember as unforgettable?
The second achievement, I think you will not believe it.

It was 1993. Hani Hussain (former Minister of oil of Kuwait, and ex-CEO of KPC) and I were discussing that as a corporation, we did not have a long term strategy. We did have a budget exercise every year, but no coherent long term strategic plan. We were thinking of a strategy that would bond all the K-companies together, the companies in the down- and up-stream of the oil sector of Kuwait under the KPC holding.

If you think about it, KPC was founded in 1980, so it was overdue time to launch our strategic plan for the group. We called it the 2020 strategy. So, we posed the necessary questions to the Ministry of Oil. What is the long term crude production target for Kuwait? To answer this we had to ask other questions such as ... What will be the production of the world in the future, for OPEC? And what will the Kuwait share be? And our spare capacity? If we knew the answer to these questions, that would guide the facilities needed to reach that vision, the number of rigs, and the workforce numbers, along with the needed technical profiles, how much gas is available for petrochemicals, how much refinery capacity do we need to build. Only then all the elements of the oil activity would fit into place.

The 2020 strategy was thought out, prepared, approved and launched. We had the platform. I was personally proud of this accomplishment, particularly because of the huge implications and benefits it will provide to the corporation.

A few weeks ago, the current leaders of KPC were talking to me about the new KPC 2040 strategic plan. I reflected that now, this is an exercise that is done every three years. So, there is a more cohesive direction for the K-Companies in the KPC holding. I am very proud.

Thursday, August the 2nd, 1990

There are some experiences in our career that mark us, just as if those are tattoos. What were those experiences for you?

Career-wise, it was to wake up on Thursday, August the 2nd, 1990, to face a different world for me, my family, my company, my country. Kuwait had been invaded.

I was based in Kuwait, but was in charge of all downstream operations worldwide, and we were holding strategy meetings in Italy. By coincidence, I stopped by in London on my way back to Kuwait, and I woke up on that dark Thursday to news that Kuwait had been invaded. All had changed.

We had to mobilize. The Ambassador in London called me the next day, and explained that the US president, George Bush, had frozen all Kuwaiti assets, to protect them from the Iraqis. There was no access to cash for Kuwaiti nationals, companies with local Kuwaiti accounts.

We had our refineries and service stations in Europe, but how do you provide crude oil to the refineries and services stations if the source of the crude has been stopped? We were in crisis management mode for more than a year.

Even if I was an EAMD at KPC, there were legal complications of executing actions and orders.

We were asked *"Who is the current authority in Kuwait?"*, *"what is your legal basis for directing ships?"* and *"What is your authority to mobilize the Bank Accounts of KPC, KOC, KNPC?"* So, as a first step we needed to make sure that KPC Board Members could meet, and could function and act as Board of Directors, directing actions for the oil sector of Kuwait.

I organized the mobilization to bring the KPC Board Members and their families to London. We placed them in apartments, and worked with them, providing the necessary legal grounds for their actions and decisions.

"Who Is the Owner?"

It was not clear to all that you owned the oil assets?
No. At least, not to all. I will give you one example. At the moment of the invasion, we had about 9 tankers offshore, delivering our crude. One of them was not responding to our requests to know its location and ETA in Europe. After investigating, it turned out that the tanker owner was trying to sell the crude in international waters, as he claimed the crude was not anymore

property of the State of Kuwait, as the country had been invaded. You see? People were trying to take advantage of the situation.

We needed Board resolutions to validate our authority in dealing with third parties.

A month later, I had a meeting with the chairman of Shell, the late Peter Holmes. Customary to his unique style, he said to me "Nader... we have been a customer of Kuwait since the sixties. How can we help you?" I told him that our refineries needed Kuwaiti crude to make lubricants,. And you have cargos coming to Europe with our crude. Holmes, who knew Shell had paid for those cargos, replied: "Take them, they are yours!" No discussion. No further conversation. Shell had bought those, but he diverted the tankers to go to our refineries in Europe.

Many companies helped us. I would say 90% of our liaisons, our clients, helped us a great deal.

The Morale on the Floor

The other part of this story is that we had to manage the morale.

The morale of the Kuwaitis in Kuwait and outside Kuwait had to be boosted back, as it was on the floor. This was the time when I became active in the public campaign *"Free Kuwait!"* we trained the speakers for the campaign, and we all had the same objective.

We hired very professional companies to lead media training, who taught us how to answer difficult questions, to face the media, and the reporters. I participated extensively. I was told: *"You can be the image representing the oil industry of Kuwait and talk about what happened on TV"*. So I ended as one of the visible images and voices of the Free Kuwait campaign in London.

We also helped the campaign by funding their advertising program.

And what was your role?

I was Deputy Managing Director of KPC, and KPI President, so I had the resources of our London office.

Planning for the Return

In London, Hani Hussain became the head of the planning effort called "Alawda", "the return" with the idea of planning how to come back to Kuwait, and regain control of our oil industry.

How long were you in London organizing the executives, the members of the KPC Board?

I was in London for 9 months, throughout all of the invasion.

The Perhaps…

But we have talked only about the good, the positive aspects. Nader, do you have any regrets?

I think every single leader has achievements, but also many challenges and some regrets as well. Challenges and regrets. One of my biggest regrets is that I wished I understood earlier and acted on how far we were behind in Kuwait in terms of the technology capability in our industry. The oil industry is heavily dependent on technology, and I realized that when I look not far away, in Aramco, they managed it very differently, and I could have done more about it. Technology should be now ingrained in our capability development.

Perhaps, I could have done a benchmarking process and insert technology as an intake, and an early adoption of technology in our approach to the operations, the studies, the plans. Our people would now be working differently.

Is there anyone who championed you? A sponsor? From who you have learned the most?

I cannot pick just one name as there were many. In my career, I have liaised with 7 Ministers of Oil. If I would mention just one, the other 5 would not be happy. But when I saw some of the Ministers in action, with their ability to chair the Board meetings, managing all the personalities you find in a board room, I gained an extraordinary insight on how to manage personalities.

I also learned a lot from Hani Hussain. His humility disarmed me. He taught me that you do not have to be aggressive to achieve your goals. I learned a lot also from Abdul Malek Al-Gharabally, and his unbeatable common sense. He once entered in my office and saw a pile of papers to be signed. He said to me; do not rush in signing the papers…"*Remember, your signature reflects your name and reputation, and not just your authority.*"

On a Scale from 1 to 10

In 1973 I had a terrible car accident in France. I had to stay in hospital for three months. As a result of that incident in which I could have died, I acquired a different perspective on life.

This new perspective has helped me all along my life, during critical moments of my career. It certainly helped me on that awful August the 2nd, 1990. I saw my family was fearing the future, due to the invasion. But I counted our blessings: we were healthy, I had a job, a safe place to stay, and there was a solution by the allies pledging to liberate our country. I did not say we did not have to fear, but in perspective, on a scale of 1–10 in terms of negative impact, the invasion for me was an 8. Not a 10. I already experienced a 10 in my life.

You also need a some relative yard stick to measure things in life. And ask yourself "*where does this rank?*" I learned this at a very young age. This is how I developed the ability to stay calm in the midst of a crisis.

How do you do to cope with the stress?

From the accident. I had a metal pin inserted in my thigh, so I went through a long period of physiotherapy. The good side of it is that I gained a love for exercise. I used to run to get rid of my stress. Now, my knees to do not let me run, but I exercise daily.

The job of an executive person is like running a marathon. You have to pace yourself, you need stamina. It is a long run, so you have to save your energy.

The "I" and the "We"

What challenges do you envision for the energy industry in the future?

I think in the short- to medium-term, we must learn how to thrive in an environment of low oil price.

Additionally, in the Kuwait context, the challenge is to handle a much more complex stakeholder environment internally and externally, with our Parliament and with a fast-changing political scenario on the global scale.

Over the longer term, we need to develop our strategies for coping with the potential impact of r Conference of Parties (COP) 21. We must try to understand the possible impact and how to mitigate the risk.

And a challenge that is very Kuwaiti-oriented, but that keeps me awake at night is coming from my interaction with young people. The values of the modern Kuwait are not those from the 60s. I noticed that the young with whom I liaise have become short-term minded, materialistic, oriented to the "I", and not the "WE". Everything is" *What do I get from it for me?*" Maintaining a set of corporate values, like professionalism, commitment, integrity and transparency in this new setting is very challenging. We need to protect the corporate values, the professionalism which is essential in the oil industry, where we are surrounded by risk.

Energy Is Not a Dirty Word

What is your message to the new generations?
Always try to get an external orientation for what you are doing. Think about adding value in terms of your customers, stakeholders, and compare yourself with your competitors.

And if you ever get depressed in the energy industry, think about the amazing value that energy brings to millions of people in poverty. Energy is not a dirty word. You can literally ensure the survival of billions of people, and you should be proud of what you are doing.

Nader, do you consider yourself to be a resilient person?
My life has been filled with challenges, which forged resilience in me. From the beginning, because, let me tell you, that going abroad to boarding school was not easy. It was tough.

I was only 11 years old, and I faced what today is called bullying. It was on my second day 3 boys started pushing me, trying to be amused by bullying newcomers to the school. They thought they were stronger, better.

But I started to laugh, and they noticed.

"Why are you laughing?" they asked. I replied *"back home, my 4 older brothers beat me every day. You guys are nothing compared to them!"* I got their respect.

The loneliness of boarding school impacted me in many ways. I did not send my three boys to boarding school. On the positive side, I learned to be very self-dependent, and it definitively forged my resilience. This helped me a lot in the corporate life.

A Shared Selfie

- **Your favorite word:** Alhamdulillah, which means "thanks God, or praise God", in Arabic. It has been my favorite word for the last twenty years.
- **A city:** Mallorca.
- **An important person of your preference:** Winston Churchill.
- **A personal happy moment:** That day, arriving at the reception desk of Gulf Oil as President of KPI, and not been recognized.
- **Your favorite food:** Mexican food. I like burritos, fajitas, tacos.

- **Your favorite color:** I do not have a favorite color. It depends on the mood of the day.
- **And your favorite music?** Depends, I like a variety of singers and songwriters in contemporary western music.
- **Who supported you the most?** My dear wife Ebtihaj.

Post-scriptum

I had to ask how come Nader liked Mexican food. He calmly explained that in 1998, on a flight of American Airlines, from Boston to Chicago, he was reading an article in the on-board magazine, featuring the "top 10 Mexican restaurants of the World", and one was in Chicago. The restaurant had a one-month waiting list for reservations, but they managed to get a table. He enjoyed the food very much, and remained a fan of Mexican food forever.

What impressed me is that he told me he went to Chicago to visit his son, studying at the University of Chicago. His son was also helping with social programs for the poor African–American communities of Chicago, in an initiative led by a very talented lady called Michelle, who was married to a young law professor teaching at the University, Barak Obama.

It is a small world indeed.

Hashem S. Hashim

"Seek out the tough roads, for they lead to amazing destinations".

A Glimpse

Kuwait Petroleum Corporation (KPC) launched a major new company in 2016 namely, the Kuwait Integrated Petrochemical Industries Company (KIPIC). KIPIC, one of the two largest companies within KPC, will integrate, deliver and run three massive new projects at the Al-Zour complex. The first of which will be a refinery, followed closely by a Liquified Natural Gas (LNG) importing facility and finally a Petrochemicals plant. The investments will total 26 billion US dollars, uplifting Kuwait's position in the downstream sector and increasing the refining capacity to 1.4 million barrels per day.

It takes a special person with exceptional skills to lead such a huge endeavor. It is therefore not a surprise that Hashem Sayed Hashim was the one chosen to be the CEO of this new company, recognizing a leader with a track record of outstanding results in the face of adversity and complex situations throughout his career within Kuwait Oil Company. A leader who loves a challenge!

From 1987, when he started as a newly hired engineer in Kuwait Oil Company (KOC), he was a natural leader. From steering the development of a tough field like Umm Gudair having complex border issues with Wafra to a total revamp of the Reservoir Management Department, and the formation of a new Jurassic Gas development asset in North Kuwait, Hashem was single minded in his commitment to the overall company objectives.

He subsequently headed the South East Kuwait asset and revived the health of Burgan by almost doubling the reserves allowing a longer term strategy to be put in place for sustained production from the maturing giant field Burgan. Building Burgan to new heights of 1.7 million barrels of oil per day and creating sound plans to sustain it, was one of the highlights of his career.

This entailed turning water disposal enhancements into water injection projects for increasing oil production. Rejuvenating aging facilities, reviving the production system whilst planning new major projects to sustain the maturing production was not easy. However a leader achieves the impossible when the right people are engaged at the right time to tackle the right issues. His sharp focus on selecting the right people has been Hashem's trademark.

When he became the CEO of KOC, he steered the company to deliver 3.0 million barrels of oil per day for the first time in the recent history of Kuwait when he created the production optimization task force and engaged the company to align as one towards its delivery in 2015. And despite sceptics at the highest level in the country, he successfully engaged IOCs (International Oil Companies) to return and support KOC for the longer term both technically but also to coach and train the next cadre of capable people to run the company.

Today he is building KIPIC from nothing with a passion and energy of a youth and wisdom of a seasoned leader with a whole new chapter to write in the coming years.

- Graduated from Kuwait University in 1987, as a Chemical Engineer.
- 1987 joins Kuwait Oil Company, and works in Projects team, testing recovery processes for the Heavy oil in Ratqa oil field.
- 1996 Superintendent, Umm Gudair (WK asset).
- 1997 Superintendent, Burgan Field Development (SEK asset).
- 1998 Superintendent, Reservoir Management.
- 1999 Head of newly created Strategic Planning Team.
- 2000 Launches the first strategic plan: the "*KOC 2020 Strategy*".
- 2002 Kuwait Project Lead—concept of Operating Service Agreement (OSA) was first created.
- 2003 Manager Field Developments South and East Kuwait (Burgan).
- 2006 Manager Jurassic Gas Development (Planning and Gas)—Initiated the first Gas Field Development in Kuwait and long term Gas Strategy.
- 2007 Deputy CEO Chief Executive Officer of South and East Kuwait (Burgan).
- 2009 SEK reaches the record production of 1.7 million barrels of oil per day, for the first time in the modern era of Kuwait.
- 2010 Signed first ETSA with IOC in Jurassic Gas.
- 2013 CEO of KOC Kuwait Oil Company. Also appointed as Chairman and Managing Director of The Kuwaiti Oil Tankers Company (KOTC).
- 2015 KOC achieves sustainable production of 3.0 million barrels of oil per day.
- 2016 KOC signs 3 major ETSA projects with two IOCs in SEK, NK Conventional and NK-Heavy Oil.
- 2016 CEO of KIPIC, Kuwait Integrated Petrochemical Industries Company.

A Personal Snapshot

The setting was amazing. The view from the 19th floor of the architecturally awarded magnificent KPC building made a perfect backdrop for this interview.

Hashem's unassuming, calm and humble style was a stark contrast from his sharp wit and passion when he talked about the projects during his career. His eyes always lit up when discussing the people in his life and how his successes were really their achievements, their dedication and their sacrifices.

An Agent of Change

Early Stories

Where were you born?

I was born in Ahmadi, as my father worked at Kuwait Oil Company and Ahmadi is the oil city that was built near the Greater Burgan oil field. I was the first son after three daughters. We lived there for many years before moving on. But I returned to Ahmadi where my children also grew up surrounded by the oil Industry.

How did you choose to study Petroleum Engineering?

Actually there was no Petroleum Engineering Department at the time I started at Kuwait University. I began by studying Medicine.

Medicine! That's very different.

Yes, Medicine. I studied for one and a half years, mostly pushed by the wishes of my family, and expectations all around me. But in my heart I wanted to follow my father's footsteps and work in the oil industry. So, I finally changed to Chemical engineering, graduated in 1987 and joined KOC where I entered a program designed to obtained a diploma in petroleum engineering.

I started in a team testing a variety of recovery processes for the Heavy Oil in the Ratqa field. I was fascinated by the technical work and quickly got into coordinating the assessment of the results obtained by screening methods at a pilot scale.

In 1988, I was transferred to the South and East Asset, and was excited to work in the giant Burgan field as a Reservoir Engineer.

But the 1990 invasion of Kuwait interrupted my reservoir engineering career. KOC workforce was volunteered to several key industries of Kuwait crucial for the country's survival. It was an opportunity to serve the nation. I worked tirelessly in the manufacturing of bread, at the national mills industry. I was assessing how to optimize the processes, identifying the value chain, and detecting if there were opportunities to improve productivity or minimize costs. It was an eye opening experience.

Wow! I do recollect being told that Kuwaitis at that time proudly bought their national brand of bread from the "Kuwait Flour Mills and Bakery Company". They remember it with reverence as one of the factories that was sustained by regular citizens in a collective volunteering effort.

Absolutely true. It was a classic example of people coming together as an unwavering will of a nation to rebuild itself.

Were you already married or with a family at the time?
No I was single and stayed in Kuwait throughout the occupation and Liberation, in our family house.

Firefighting

Tell us about the experience towards the end of the invasion and the start of the reconstruction of the country?
As we all know the invasion ended with all our wells in Kuwait on fire. It was horrific and unimaginable. I was part of the wells firefighting team, supporting technically and informing the configuration of the wells, completion, depths etc.; providing all the information necessary to allow the wells to be controlled and the fires extinguished.

Almost immediately, we had the monumental task of returning the production back to normal. It was such a hectic period, with a high level of simultaneous activity in drilling, workover and reconnections all over the fields of Kuwait. Our country needed us to boost our production and capacity to meet 2 million barrels per day.

I resumed my role as a Reservoir Engineer with a difference. The experience of reviving our oil fields after destruction and devastation was truly unique. It was like 10 years of learning packed into one year. All of us at the time grew beyond our youthful years. And even after the last fire was put out, the stench in the air stayed with us for months whilst the ground pollution stayed for many years. We are still cleaning up some of the environmental damage 26 years on.

Once we restored the production to 2 million, KOC developed a longer-term strategy for growth, the first one in its entire history. We envisioned an aggressive target capacity of 3 million barrels of oil per day by 2010, preserving Burgan to be a swing producer and focusing on developing new and difficult areas first. We had large reserves which were yet to be exploited to secure the future of the country as a strong oil producer. We successfully reached that peak capacity in October 2010. The next challenge was to create a sustainable production at these levels by 2015 which was also achieved against many challenges from water handling and water injection projects.

Your commitment to your country is truly remarkable. Did you ever have the opportunity to work overseas?
Yes. I was assigned to work with BP in Aberdeen, on their offshore operations in the Miller field, North Sea. It was yet another diverse learning experience for me. I discovered the true value of integrated working.

I understood how a lean, profit focused company makes every decision by weighing costs against return on investment. And how these conversations were taking place across the whole company to ensure people were fully aligned to consider profitability in every action they take.

Transition from Extracting Oil in Easy Sands to Tougher Carbonates

As insights for the new generation, what can you tell us about your earlier journey that shaped your career?

When the company transferred me from Burgan Asset to West Kuwait Asset in 1995, it was a difficult move at first. I was too comfortable working with sandstones and knew very little about carbonates. As I was to learn, we never truly grow until we are thrown out of our comfort zone. I was only a young professional, but was presented with many issues that I volunteered to resolve. I put my hand up more than anybody else to take on difficult tasks. It wasn't long before my talents were recognized and I was promoted to be Superintendent of Fields Development for "*Umm Gudair*" carbonate oil field. One of the initiatives I led was to control the oil migration process which was losing reserves for KOC, with the "Line of Defense" concept that saved the company billions of dollars in future revenue.

I moved back to South and East Kuwait as Team Leader of a special initiative called "*Modelling of Burgan*". It was a huge modeling project of the subsurface, to be carried out in KOC with Chevron. The size of the model was mind blowing considering Burgan is the largest clastic reservoir in the world. Our second challenge was difficulty in enhancing production due to pressure decline in Wara, the second largest reservoir in Burgan. In order to implement the concept of water injection for Wara, we needed to pilot the concept quite quickly so as not to delay the first large scale injection project. The third challenge was to develop people's skills in how to manage our fields facing increased water cuts and with new demands on managing a massive waterflood project.

Looking back at these enormous challenges and how we tackled them, I can reflect that they were the start of a big change in the development direction of Burgan to create new value for the company.

In 1998 I was thrown into a completely different kind of challenge. Appointed as Team Leader of Reservoir Management, I observed that the department had too many functions, diluted across vastly different scopes and urgently in need of more focus and clarity in direction. I re-organized the whole group sending IT-related material to a new team thus, launching the KOC Data Management department. Reservoir Management now has the companywide accountability to audit the necessary processes that guarantee the health of our reservoirs, share sound reservoir management practices to assure sustainable production for generations to come.

A Companywide Perspective

In 1999, a new team called Strategic Planning Team was established and I was appointed as the Head of the team. My assignment was to create the 2020 strategy for KOC and to develop the right processes and organization that will deliver it.

What a fantastic assignment!

Yes, the strategic planning experience early in my career truly provided the broadest perspective on how our company operates and delivers a sustainable future for the country's treasures.

What happened next?

In 2002, I was promoted to become the Technical Manager, for "*Kuwait Project*". KOC observed that TSA Technical Services Agreements with International Oil Companies (IOCs) did not provide enough incentives or scope to develop both the national workforce and our oil fields effectively. A fresh look at the contractual structure was needed. The focus of "Kuwait Project" and my assignment was to develop a new model for collaboration with IOCs. I led a joint team across KOC, to developed a concept that would evolve into the Operating Services Agreements (OSAs) with IOCs.

After that, 2003 saw me moving back to South and East Kuwait (SEK) as a Manager of Field Development. We were at a critical stage of upgrading the gathering centers to prepare for the growth in production of 1.7 million of barrels of oil. I aligned all plans and activities and streamlined the pipeline of projects, avoiding much duplication.

I saved the company a few hundred million dollars by stopping new disposal plants and combining the disposal concept with an upcoming injection project. This opportunity for efficiency was not seen because of lack of integration across the group at the time. Even the Deputy Managing Director at the time found it difficult to handle the cancelling of this project. I stepped up to take responsibility for the change. Looking for integration and alignment in activities and plans across the whole asset became my theme.

A Major Step into Gas for KOC—A Tough Challenge for Me

What was the moment of significant change for you?

In 2006 KOC announced the discovery of commercial quantities of Jurassic Gas in North Kuwait. Creation of a new Gas Strategy and an aggressive plan to develop it landed on my lap as I was appointed Manager of Field Development of Jurassic. Being at the start of a new era in KOC's Gas strategy allowed me the opportunity to make this new project truly multi-disciplinary and fully integrated with the right people and teams to ensure success. Developing Deep Sour Gas was not the easiest way to start a new Gas Development Strategy in Kuwait. This was one of the toughest challenges I faced in my early to mid-life career.

Almost Doubled the Reserves of Greater Burgan

Greater Burgan is a major player in the world economy. What transformations happened after your return?

In 2007 I returned to SEK as Deputy Managing Director of the asset. My objective was to reduce project cycle-time for business critical upgrades. I directed a major facilities' upgrade and drilling program in Burgan Field including engineering modifications to 14 production Gathering Centers (GCs). This was accomplished within 3 years, whereas a typical KOC project cycle would have been 6 years, without any interruption to production whilst maintaining the highest standards of HSSE (Health, Safety, Security and the Environment).

Another major change was turning the Chevron-driven pattern waterflood concept in Wara to a lower risk peripheral waterflood. Another key step towards risk reduction was a phase one injection scheme that I initiated called Early Wara PMP. This project was an amazing example of people being engaged and fired-up to deliver excellence. The project delivered in a record breaking time of 6 months from sanction to injection start and the early learnings from the waterflood saved a few hundred million dollars in the main Wara PMP waterflood. I still recall the 6 months of buzz and the passion in the people who made it happen across all disciplines within and outside the asset including our service company partners. This record is yet to be broken for a similar size project in KOC. It truly helped me understand how much power is possible from people's untapped potential.

Another massive change that influenced our strategy for KOC, KPC and the country was the discovery of hidden potential in our giant reservoirs that almost doubled the remaining reserves in Burgan in 2010. I guided the study that led to this discovery providing the challenge needed to extract an outstanding outcome. This set the scene for a whole new development strategy for Burgan where we created a Life of Field Strategy, the first in the history of Greater Burgan Field. The one that sparked similar life of field strategies to be put in place for all the fields in KOC. We also reached the target of demonstrating a producing capacity of 1.7 million barrels of oil per day in 2010. It was the first time Burgan produced this volume in the modern era of Kuwait.

I can see you are an agent of change....

Change for good reasons. There are many stories where I was simply the spark and a guiding light that removed barriers in the system and allowed my people to run unrestricted towards their goals. It is not enough to provide focus and clarity in direction, sometimes as leaders we need to intervene to remove obstacles to progress.

Ascension to CEO

Now, tell us about the moment you were called into become CEO of KOC. What did you feel at that time?

KOC's role in the oil sector is very challenging and difficult, in the current politically complex environment. Leading KOC in the 70s or 80s was different than it is today, as KOC's influence in the oil sector of Kuwait has undergone many changes.

When Sami Al-Rushaid wanted to step down, the KPC CEO called me to replace him as I was the only suitable leader at that time to take this position. I questioned my suitability as my focus has always been more on technical and business aspects of the company and less on responding to the frequently changing political environment in Kuwait. I stated my reservations, but there was no way out and I was appointed.

It was clear to me at that very moment that if my focus remained fully on expediting the technical and business objectives of the company, my tenure as a CEO of KOC may not be a long one. The beauty of our democracy is that change can be sudden and bring new people with fresh ideas to take up positions in government to drive a new agenda for the country. The downside is that if these changes are too frequent, it takes time to get the wheels in motion again to progress key investment decisions on projects.

Some investment programs can be deferred without consequences. However, in the upstream oil and gas sector, pace of decisions is critical to the successful implementation of our long term production strategy and many projects are very time sensitive. To drive the pace of decisions required to avoid future gaps in oil production and to provide a sustainable income from the biggest revenue generator in our country, I would have to take personal risks. I decided I was willing to make that sacrifice.

What has been your biggest challenge so far?

I knew the CEO role would bring new and exciting opportunities to shape up the company with significant challenges along its journey. But I probably only grasped the real size of the responsibility when I had to lead not only the businesses associated with oil production growth, which were already extremely complex, but also those projects associated with numerous other supporting sectors.

Aside from mega investment projects in oil expansion needing urgent decisions, there were a multitude of other expansion projects in the pipeline that also needed clear direction to close out issues and progress. Projects like the new Ahmadi Hospital, expanding the marine capability, growing the Ahmadi Township, rapid expansions of Power stations, which require swift expansion in the fuel network to meet the increased demands on electricity consumption in Kuwait, to name just a few!

Protecting our heritage and pride in our Ahmadi township required attention to projects that restored the formal beauty of the place before the war. Small projects such as changing out unattractive metal fences introduced after the gulf war as part of quick repairs to the township, to the original style of wooden ones made a big difference to the environment of our oil heritage. It's the small things in life that can lift our moral giving us strength to face bigger challenges. Adding to this I faced changes from the new laws that banned professionals with tenures over 35 years to stay in the company. This meant 40% of the senior management needed to change out and have new appointees. This was the first time that a CEO had to manage 9 freshly appointed deputy CEOs, along with a 45% change in Managers, and 40% change in Team Leaders. This was the first time in the history of the oil Industry of Kuwait that we have seen such a transformation in leadership and risked continuity issues. Despite these issues, I found ways to unite the leadership under the idea of ONE KOC and set myself to steer the company to success.

The Pearls of Knowledge and Experience

It is inspiring to see how you managed the company despite these harsh conditions. People in the company rarely understand the extent of the pressures on senior leadership.

This was just the start of what I faced. There was a legacy of major investment plans that were pending resolution and decisions. These were tough decisions to make quickly and required huge effort from me and my team to unravel all the blocks to progress. The complexity lays in the interlinks between projects, the risks of duplication, the opportunity to integrate to gain efficiencies versus delay from waiting on integration issues. One by one through a very uphill struggle we closed out countless issues and created a new direction for the company. For example, the challenges of our maturing fields and aggressive growth agenda required a substantial increase in rigs from 30 to 100 in a very short timeframe. That was a 300% increase in activity in a complex operating environment. It was impossible to sanction without the corresponding transformation in the organization and support systems throughout KOC to be able to manage the increased activity effectively. Thanks to the support of my leadership team at every level in the company, we achieved this massive feat.

Another two issues that needed urgent resolution was pace of recruitment to replace the sudden reduction of staff that reached 35 years of experience and the increasingly young workforce with skills and capability gaps. Our first achievement was in successfully increasing graduate intake from 200/year to 800/year. I am really proud to welcome more of our youngsters into the oil industry.

The next issue was to expand the inclusion of IOCs in our company to increase the level of capability through close co-operation, on the job coaching and international work experience assignments. A pilot model in 2010 using ETSA principles worked very well in the Jurassic gas project. However significant challenge from audit and legal had stifled progress on expanding this model to other assets. We were stuck. It took enormous effort from me and my team to face all the questioning in court, in parliament, engaging with KPC, Oil ministry etc. Before we could close out successfully all the mistaken charges raised against this ETSA. An important aspect of fighting this battle was to shield the rest of KOC from the distraction. That was my role; to act as a buffer, a safeguard to protect KOC from interruptions in driving forward and delivering on production promises. By taking total accountability I also deflected the legal heat away from my people. This was a big drain on our limited resources but we did not give up and the fruits of this effort came when we successfully engaged two major IOCs to support our largest Oil and Gas assets, SEK and North Kuwait in 2016. It is now up to the next generations in KOC to squeeze value out of this amazing opportunity.

And it will not be easy; it is never easy diving for the pearls of knowledge and experience. I recall my time working with IOCs under TSA after the invasion. I absorbed, questioned and challenged everything in pursuit of learning from the best. I put my hand up to participate in any joint projects so that I would sharpen my skills and deepen my experience. Although the journey was tough for the multiple teams throughout the organization who raised their bar to meet these challenges and support me, the outcomes from our joint efforts were truly amazing. After an intense and personally rewarding experience in KOC, it was a welcomed change to move out of the 'hottest seat' in upstream and dive into the business of the downstream challenge in KIPIC.

From Upstream to Downstream—Breaking New Ground in Creating KIPIC

Please tell us about the new assignment, as KIPIC CEO. Not everyone has the energy to start a whole new company as big as this one!

I do feel this renewed energy within me. People in high places have placed full trust in me to establish the company and deliver its strategy and I will not disappoint them.

How do you feel about working in downstream, after a full career dedicated to upstream?

I am a process engineer and processing in downstream needs different experience from upstream, however at the CEO level, the leadership skills required are the same. Experiencing change every two to three years has been tough, however it provided me with the opportunity to broaden my experience and mature sufficiently as a leader to be able to steer any company in the oil and gas arena.

Every three years! These are clear messages for the next generations. It is early days yet in KIPIC, but can you share any new insights?

Starting a brand new company from nothing and taking over 3 ongoing projects from KNPC and PIC brings a whole new meaning to the word 'challenging'. I was faced with 2 mega tasks. The first was starting a new mega company with all its legal, financial, organizational and contractual frameworks. When was the last time we did this in KPC? Historically the largest revenue generating companies in KPC have been ready-made, well established 'take-overs' from International companies who set them up originally.

Although this was a multifaceted task with such complexities, never before undertaken in KPC, the pace of starting a new company is less critical when there are no projects already initiated. But this is not what I faced. The next mega task was to transfer 3 ongoing massive projects worth 26 billion dollars, two of which were critically in the construction phase. Imagine the management of change needed to transfer multi-billion dollar projects without disruptions to the work. Imagine, the legal and contractual nightmare to ensure full accountability and liability issues were intact with existing contracts moving across from one company to another brand-new entity. Imagine the risks to project delivery schedules if anything was over-looked. And imagine the legal nightmare from contractor claims if a minor detail or two was missed.

I also face misconceived ideas on the pace of the transition. This was due to the lack of experience on the complexities involved in setting up KIPIC, within KPC and the ministry—they thought it would be easy and quick. It wasn't easy, but despite many obstacles, we delivered at top speed with people working all hours to make it happen. All the protocols are now complete to legalize the movement of the assets from KNPC without any disruptions to the projects. A major hurdle and a major milestone reached successfully. The journey has just started and the first chapter in the history of KIPIC is being written. The most important next step is happening—right people in the right places within the organization to deal with the diverse demands of the 3 mega projects, the Refinery, the LNG plant and the Petrochemicals all integrated in one site at Al-Zour—this kind of cross function integration downstream is also a first for a K-company.

It Is All About the Team

What can you share about your leadership style?

I can be quite persistent and unwavering in pursuing what I believe is good for the company. However, my biggest strengths lie in challenging the norms, in breaking down barriers and more importantly in hearing out new ideas no matter how unconventional. I rarely take 'no' or 'not possible' for an answer. They say if you do what you've always done you will get what you've always got.

How can we expect a different outcome if we are not willing to do things differently? I am open to new ways of working, willing to take risks and persevere where others give up. I also take care to give a lot of space for talented people to deliver and provide timely support when needed. My strength lies in being able to engage individuals to succeed. Our people are the building blocks of our successes and their true power is rarely unleashed fully until they become part of the decision-making process. Business transformation comes as much from within as from importing best practices from around the world. How we do business in the rest of the world is not always directly relevant to Kuwait. There is a strong element of people and culture that needs to be blended and my leadership style lies in creating the right balance between these elements to tailor solutions that are locally effective in Kuwait.

Renewed Longevity of Greater Burgan, Protecting HSSE and Uniting KOC to Deliver on a Recent, Huge Production Challenge

Where do you consider your leadership added most value?

During my career there were many occasions that left a mark on me and I would like to expand on 4 that occurred in the last 10 years.

Achieving the discovery of billions of barrels of new reserves in our maturing field—Greater Burgan was not a chance event. I had a gut feeling that there was more to Burgan than we were led to believe and I followed through with this instinct that led to the discovery. This gave us the opportunity to make significant new development plans to access these reserves and increase our aspirations towards a longer sustained future. These new reserves and investment plans, added previously unimaginable value. The new value created went into in hundreds of billions of dollars in net present value at 7%, which was equivalent at the time in 2012, to the market capitalization of the largest private oil and gas company in the world—Exxon Mobil.

To be part of creating a whole new future for our next generations is humbling. I felt tremendous pride in the people within my team who made this happen. To have given our country's largest asset a whole new lease of life is an unforgettable event in my career and I hope in theirs too. We have made history together. These are stories we will be telling our grandchildren and great grandchildren in years to come.

The message is not to ignore your gut feeling, your instincts.

The next 2 events that changed me are to do with people and safety. Protecting people and creating a safe environment to work and live in, has always been a priority with me.

I led a complex HSSE incident in the Ahmadi township following gas leaks that caused explosions. After rapid evacuation of part of the township, novel solutions were implemented to monitor, mitigate and sustain its containment, managing delicate interfaces with government and environmental bodies in the process. The next event was the recovery from a major HSSE crisis when an exploration drilling into a new opportunity resulted in a poisonous sour gas blow-out. I led the successful containment of the well with no harm to people and environment breaking new ground in emergency response standards in KOC.

These two incidents changed me as they had the potential to cause significant damage and disruption to the lives of many people and their families. I look back and feel that there must have been a guiding hand that helped us contain these potential disasters. More importantly it taught me not to take for granted the fragile nature of living close to our giant oil and gas reserves. And that our biggest treasure is our people, not the liquid gold from the ground.

Wow! People rarely see how a potential major disaster was averted from a small incident

The last major event I recall was the most recent challenge in 2015 when production ceased in the neutral zone. I was the CEO of KOC at the time. We were stretched on meeting our production targets. So how do you create something out of nothing at such short notice? And yet, we did pull a hat-trick and added production to offset the deficit within a record 10 months. This was no magic. This was the power of people uniting behind one goal, one purpose with unwavering commitment to the success of KOC.

It required aligning the senior leadership and managers to work across the organization, demolishing all boundaries. Setting rules to make this happen was tough. Creating space for assets to forgo their individual KPIs to meet the company objective was not easy. Appointing a deputy CEO (DCEO) of one asset as lead for this task, to interfere in the business of other assets required humility and sacrifice on the part of the other DCEOs.

To my great delight I saw people stepping up and helping across assets with allegiance to just KOC as a whole, not the team nor the asset that they came from. I look back with pride at how by providing the right framework, the right support and believing in the capabilities of my team, the impossible became possible. I see those same people driving outstanding performance today in KOC and I continue to be very proud of them. As I watch them deliver quietly and humbly behind the scenes, I say to myself these are our leaders of tomorrow!

Alternative Energy Sources Are Here to Stay—Get Onboard

What are the challenges that you see for the energy industry in the future?

The energy industry is making increased headway towards renewables by driving down the costs to be competitive with conventional. What does this mean for Kuwait? We need to reassess the diversification in our local energy consumption and start planning for the longer term. If it is more cost efficient to have renewable and greener energy then it allows us to preserve our depleting oil and gas reserves and extend the life of its supply for our needs. And it could also present an economically attractive opportunity to export more by transitioning the local consumption to renewables.

Another challenge comes along with the new opportunities of reaping efficiencies from becoming fully digital. The world needs to be better prepared against the increasing risks of cyberattacks. In Kuwait we need to be simultaneously prepared on cyber security as our assets move towards the opportunities of a fully digital operating environment. And although coordinating inter-related diverse projects is an ongoing challenge, we cannot stop the digital progress if we are to be competitive on development and extraction costs.

A new challenge is emerging from the recent downturn in oil prices. It has created a different mindset worldwide and capital investment has moved from long term complex major projects to short cycle time, smaller investments with faster returns. This change in appetite towards lower risk projects cannot sustain a long term supply in the oil and gas industry. However, this challenge for the rest of the world in the coming years could be an opportunity for the OPEC suppliers. So, the message for Kuwait is to be ready—be cost efficient and competitive with the rest of the world compared to alternate energy sources, embrace the opportunities of renewables that could thrive in our environment, be smarter with digital technology and be proactive on cybersecurity.

Take Control of Shaping Your Future Today—Walk the Talk

What message would you like to send to the young generation?

I have huge expectations from the youth of today as they have a monumental task to absorb knowledge and experience from those about to retire from the energy sector. Tomorrow's energy industry is increasingly about best-in-class and performance excellence. It is about surviving in a very competitive environment and showcasing your ability to tackle tougher challenges in maturing reservoirs.

It is about thinking bigger than the team, group and asset you are in. It is about demonstrating early in your career that you can step up to lead by example not by instruction. I repeat, 'lead by example' —walk the talk. It is about showing that you are able to engage and motivate people to deliver. It is about using the carrot far more than the stick to galvanize teams into action.

Tomorrow's great leaders will have admirers very early in their careers and respected for the way they conduct business with professionalism and in how they lead teams with humility.

We tend to focus a lot on technical coaching and my advice to the young generation is to seek people who will also coach and advise them to be leaders of tomorrow. Find your true passion in the kind of work you choose. Find a role model to inspire you and find a mentor to guide you. Ask yourself what the future leadership in the K companies would look like for the new challenges ahead? And how can you play a part in shaping that future? Ask yourself if you really need anybody's permission or invitation to be a leader in any field that fires your interest? If you see tomorrow's future has to be very different from what you see today then start with the person that you can influence the most—yourself. Be the change you wish to see, live tomorrow's future today.

A Shared Selfie

- **Your favorite word:** Value.
- **A city:** Ahmadi, Kuwait, the city I was born.
- **An important person for you:** My mother.

- **Your favorite food:** A well-prepared Zubaidi, it is my favorite Kuwaiti dish.
- **Who supported you the most?** My mother.

Sami Fahed Al-Rushaid

"We cannot shy away from challenges".

A Glimpse

The State of Kuwait has produced oil and gas for over 70 years. And it has also produced an extraordinary lineage of leaders to steer the destiny of its oil and gas industry.

One of the individuals who has given to Kuwait a special dazzling light in leadership is Sami Al-Rushaid. Al-Rushaid steadily rose to the heights of the two most important companies of the Kuwait Petroleum Corporation: Kuwait Oil Company (KOC), in the upstream segment; and Kuwait National Petroleum Company (KNPC), in the downstream one. Few other leaders have attained the same experience, which provides him with an extraordinary viewpoint. From his perspective, the whole country is in the frame, and the employees are the enablers of the future of Kuwait.

The contributions of Al-Rushaid to his country are numerous and cherished by his colleagues. A prime promoter of the women empowerment in Kuwait, Al-Rushaid was the first endorser of the "Professional Women Network" of KOC, an initiative founded with his firm support back in 2009 by Hosnia Hashim, when not many organizations or sectors were talking about diversity.

Accustomed to steering the destinies of big corporations, Sami Al-Rushaid is now a Member of the Board of Directors of Kuwait Petroleum Corporation. The leadership of the oil sector of his country did not release him upon his retirement, and invited him to keep contributing for the future of one of the most vibrant industries in the world.

- B.S. in Industrial Engineering, University of Miami, 1978.
- In 1978, joins Kuwait National Petroleum Company (KNPC) as an assistant planner.
- 1987, KNPC Plans Coordination Manager.
- 1995, Executive Assistant Managing Director (EAMD) for Planning and Finance KNPC.
- 1999, Deputy Chairman and Executive Assistant Managing Director (EAMD) for Planning and Projects, KNPC.
- 2000, Deputy Chairman and EAMD for Manufacturing, KNPC.
- 2004, Kuwait National Petroleum Company (KNPC), Chairman and Managing Director, responsible for Kuwait's in-country refineries, Al-Ahmadi, Shuaiba, and Abdulla.
- 2007, Kuwait Oil Company (KOC), Chairman and Managing Director, responsible for oil exploration and production operations in Kuwait.
- 2014, Chair and Board member of several organizations, including Kuwait Scientific Center, Warba Bank, and GEMS.
- 2017 Kuwait Petroleum Corporation (KPC), Board Member.

A Personal Snapshot

Sami Al-Rushaid may be defined as a progressive leader. He is a leader who has transformed his leadership style and perspective through experience. In his own words, he developed his leadership gradually, in one extraordinary journey that is a compilation of talent with opportunity and overall, an unrivaled tenacity.

Sami Al-Rushaid has been a colleague of Hosnia for many years, in the leadership of the oil industry of Kuwait. He was the Chairman and Managing Director of KOC when Maria joined the company. Our relation with Sami is a fruitful one, which especially benefits the young talent of Kuwait. At the beginning, it involved sharing his insights with the high school students of Kuwait, in the "Energy-4-Me" outreach programs of the Society of Petroleum Engineers (SPE), and now sharing his vision and learnings about resilience and leadership here with us.

The candid and open way with which he approached our request to include his journey of success in our compilation was motivating. His story, told in his own words, made us marvel once more about the direct and humble approach of Sami Al-Rushaid to life and work.

Arranging the Interview

Contacting professional colleagues who have finished their formal work commitment, and are engaged in a "retirement" phase of life, would seem to be an easy thing to do. But not with Al-Rushaid, because he is as busy as ever.

With the affability and pleasantness that characterizes him, we agreed on a meeting. The conversation was a continuous flow of findings and discoveries. The quiet ways of Al-Rushaid enabled a peek into a wonderful journey towards the heights of the leadership of the Kuwait oil industry.

I Evolved from 'Analytical' to 'Driver'

Where did you study elementary and high school? Were you a good student?

I studied in Kuwait. I would say that yes, I was a good student. But I must admit to you that my high school period was one in which I was mostly dedicated to Sports.

Oh! I would have never thought you were into sports. Tell me more!

I used to be very good at Volleyball. I played for the Kuwait Men's League, in the AlQadsia Volleyball Club, one of the top nine volleyball clubs in Kuwait. AlQadsia, was always a finalist in the Kuwait Volleyball league, and we often reached first or second place in the yearly championships.

Champion at sports.

You may say that I was a champion at sports. I trained extensively, focusing on and excelling at volleyball. I trained every single day, and as I did not drive, I took the bus to go back and forth to the training. Four hours of daily training and commuting, arriving home exhausted but happy. I did this for the last three years in high school.

And the studies? What career did you want to study?

With such a strong interest and time dedicated to sports, my grades during high school were fairly average. But I had one clear objective in mind, and that was to study engineering. At the time, the Kuwait education system offered scholarships for the best students interested in engineering, to study in universities in the USA. The selection for scholarships were based only on the results of a final examination during the 12th grade of high school. So, if you did well, you would have an opportunity; if not, that opportunity was lost.

I informed my volleyball club that I could not participate in any matches of the league activities, because during the last 45 days before my high school final examination, I would dedicate my time only to studying. I hired two private instructors: one for math, and for physics, and studied day and night for 45 days. I prepared a strict study schedule that I followed without derailing from it to the end.

This effort paid off. I made the percentage required for the scholarship!

Do you remember your score?

It was a peak moment in my life. How could I forget?! I did extremely well in math, the maximum score was 120, and I obtained 117. It was the summer of 1973.

Sunny Weather

So, you had to select which American University you would have preferred to attend.

Yes, and that decision comes as a result of an interesting story. The first phase of the scholarship was a training program in the English language. In the USA, the Kuwait Embassy's Cultural Division, assigned me to Florida. I was certain I did not want to go to a cold place, so I decided to remain in

Florida for my studies, and picked West Palm Beach, in Florida State, where I stayed for my entire education. I attended the University of Miami, and graduated as an Industrial Engineer in 1978.

The Oil Industry

When I graduated, I was offered the opportunity to enter in the oil industry by two companies: the Petrochemicals Industries Company (PIC), and the Kuwait National Petroleum Corporation (KNPC), the company in charge of the national refineries in Kuwait. I knew a person in the Human Resources function, who signaled to me that KNPC was a bigger company, and that perhaps would offer a richer career path for me, as an industrial engineer. So, I chose KNPC.

How did you started your career in KNPC?

I was hired immediately after finalizing my studies at the University of Miami. In KNPC, I was assigned to the Planning department. I was in the first batch of fresh graduates to join the Planning department. This is not feasible anymore, and experience is required to join the Planning departments in the Kuwait oil sector companies. I evolved with different activities, when I had to interrupt my work to go for military service. It was compulsory in Kuwait, and it created a gap of a full year in my career.

I returned to the Planning department in KNPC after my military service. But my progression into leadership roles had still to wait for an incident that would be a life changer for me.

The Opportunity of a Lifetime

A few years after my return to the planning department, my direct manager informed me that the Division Manager had suddenly passed away.

We were only few Kuwaiti nationals in the Planning Department at the time, as all the other employees were expatriates. The deceased Manager was holding two roles before his demise, which was in many ways unique in the industry: Corporate Planning Manager and at the same time Deputy Managing Director (DMD) for Planning. Suddenly, there was an urgent and unplanned need to fill-in the position of DMD of Planning at Kuwait Petroleum Corporation.

They decided to appoint me as Corporate Planning Manager, which was a tremendous promotion, and a leap of confidence from them, as I had not had supervisory experience yet. Looking back in retrospect, I am sure that the

KNPC management must have had doubts about appointing me, but I know from direct conversations with some of them, later on, that their doubts were less of a concern than their reluctance to bring someone from another department for the role.

This was the opportunity of a lifetime. It was 1987, and suddenly, for the first time I felt what is a constant in leadership: the excitement and worry together. One along the other, thrill and concern, always paired.

From a new hire to a Manager position in 9 years!

Yes, I joined KNPC in 1978, and they graced me with this post in 1987. My official role was "Plans Coordination manager and Acting Corporate Planning Manager". This encompassed the three refineries in Kuwait: Mina Al-Ahmadi Refinery, Shuaiba Refinery, Mina Abdulla Refinery. It was a leap in responsibility of incommensurable dimensions.

Reading Before Signing

It was a big challenge, and one that I had not experienced before. As is my style, I did not take things lightly. I applied some elements learned from the previous Manager who held that role.

He was an elderly Egyptian, and was very experienced in planning. He was a man of details, who would not sign any paper that he had not read or reviewed. I picked this from him. I do not sign any paper that I have not read and understood in detail. If I am convinced once I've read the document, I will not only sign, but also support and champion what I have signed.

In my new role, I had a very good, technical and unconditional support from all in the Planning department, and of course, from my peers in management.

A Key Mentorship

I want to name somebody to whom I feel I owe a great deal in my foray into leadership. This person is Khalid Bo Hamra, who was at the time DMD of Manufacturing of KNPC. In his role he handled Planning and Manufacturing, so as Manager Corporate Planning, I reported directly to him. The refinery managers were very experienced and senior, whereas I was the youngest of the leadership team, with scarce experience. I received decisive support that propelled me into an integrated style of leadership, with a perspective on all functions of KNPC.

The relationship between the refineries management and the planning department, which I inherited, was a tense one. I guess is the same in every industry. Operations and Planning are at different ends of the same equation, and to reconcile the objectives of these two ends may create friction. The planning role is similar to that of an auditor, so, no operations or—in my case—refinery managers do not necessarily welcome it. This was gradually solved, as I consider I brought a fresh approach to the liaison. Having to learn and accommodate quickly to the new management role, my peers in management in KNPC transformed what was peer-antagonism into mentorship for a young colleague. It was a win-win experience.

Additionally, the DMD Manufacturing, Bo Hamra, in his great professionalism and insight, mentored me in a direct way. This boosted my skills, my perspective, and my own insight, to quickly grow to the role I had to fulfill.

The Hidden Jewels

Some of the old generation of managers would provide me with a feedback I will not forget: that KNPC had hidden jewels that did not come to light unless dug out, cut, and exposed to the sun, in reference to me. That flattered me, but most importantly, this learning served me in my career. I have provided opportunities of great responsibility to young professionals who in turn have matched or surpassed the objectives set for them. I expanded what was done with me further.

You have to take the risk, as a leader, to place young professionals in challenging roles. But it is generally a calculated risk. Especially if you pick a serious professional. They will grow into the role, and you will have gained an additional "jewel" for your team. In my case, I proved myself.

Did you realize during those years that you had a strong leadership within you?

That is a very good question indeed, as it makes me reflect.

(Sami pauses a little bit, I can see he is reflecting. I waited, forcedly showing a relaxed attitude I did not feel inside me. I wanted to know! When does a giant leader feel he is one? I was surprised by his answer.)

I must say that no, that still, the inner consciousness about my leadership was not there. I was focused on doing my job the best I could. I was focused on delivering.

Cherry-Picked to Lead!

I was promoted to Manager in 1987. Then, on August 2, 1990 the invasion of Kuwait occurred. It wasn't until 1992 that Kuwait returned to normal functionality. All K-companies subsidiaries of Kuwait Petroleum Corporation (KPC), gradually restarted normal operations.

Throughout the invasion and in the aftermath, I maintained my role as Manager Planning. In this capacity, I was always very close to the leadership of the company. Key strategic elements for KNPC, like forecast plans, budget, facilities expansion projects, and many more, would always come to the leadership directly from Planning, and they were delivered by me. I became an incrementally influential and key component of the leadership team.

In 1995, I was promoted to a directorship role, as Executive Assistant Managing Director for Planning and Finance of KNPC.

The Call of Leadership

Did you know financial analysis and finances at the time of your promotion?

I would respond to you with a big "yes", because I was already heavily involved in financials, as part of my work with the Planning Department, assessing budgetary requirements of our CAPEX and OPEX, for the short and long term. It is during these years that I discovered the leadership in me.

Tell me about this, how did you become aware of your leadership?

It was a gradual process, an inner realization.

I was leading and had led many activities, but always thinking about those as tasks. But when the appointment of DMD arrived, I felt I was ready for it, and that I had a vision and a plan in mind, which I wanted to execute. Gradually, I internalized that all these elements that composed my leadership. I had become a leader.

Who helped you the most?

I am always very serious in my work. And of course, the role of DMD required dedication. I must recognize at this point that I could be dedicated to my work in full, due to the strong support of my wife, who at her end was completely taking care of the family matters. It is she who enabled me with the freedom to dedicate all my time to my work.

As Executive Assistant Managing Director, I grew in accountabilities, vision and experience. In those years, I regained contact with an individual who was to become another great mentor and role model in my professional life, Mr. Hani Hussain, former Minister of Oil of Kuwait. Mr. Hussain was appointed Chairman and Managing Director of KNPC in 1998, when I was EMDM for Planning & Finances, and he is who promoted me in 1999, to Deputy Chairman and EAMD Planning and Projects. I was one echelon higher in the leadership staircase.

Projects was a new area for me, and I tackled it with the seriousness and methodologic approach I like. I was in this role for just over a year. Another fortuitous event was to occur that shaped my destiny again.

June 25, 2000

On the early morning hours of the 25th of June, 2000, a violent explosion affected the Ahmadi refinery. It was one of the fiercest fires in the history of Kuwait refineries, causing five fatalities and injuring 49 people, in what was a horrible incident with severe repercussions in the oil and gas, as well as in the political sector. Two units with combined capacity of 18,000 barrels per day of gasoline were destroyed, and a 120,000 barrel distillation unit was damaged.

A week earlier, in an unrelated incident, two Kuwaiti KNPC employees, an engineer and a supervisor of operations, died in hydrogen sulphide gas leakage incident in a reactor, in the other refinery located in Shuaiba,

These two major incidents in the downstream sector of Kuwait attracted the severe criticism of the political sector in Kuwait, and our Chairman and Managing Director presented his resignation to the Minister of Oil of Kuwait, which was not approved.

Within KNPC, the situation was not easy neither, and the climate within the leadership was rarified. There were logical serious questions in the public opinion, voiced out in the National Parliament of Kuwait, and then in the media: *"Why did the two major accidents happen?"* And ...*"Why do we still have the same people in KNPC?"* These were the questions that were the order of the day. They were obviously targeted at promoting a renewal in the oil sector leadership, seeking different leaders who would ensure and implement preventive solutions, as this could not happen again.

These incidents shook the company upside down and across. It was a crisis in every sense. No one wanted the role of DMD of Manufacturing, the role in charge of operations of all refineries.

I volunteered for it.

In the worst moment in the history of the refineries of Kuwait? That certainly requires courage.

The Chairman, Hani Hussain, appreciated my offer and did not answer immediately. After a few days of consultation with the extended senior management, and with the top ranks at KPC, the Ministry of Oil and the Government, he accepted, and I was appointed Deputy Chairman and EAMD for Manufacturing, KNPC, in 2000.

The Top Priority

It was really a difficult moment. I cannot describe adequately to you how our people felt. Just think: we all experienced an industrial tragedy. Dear colleagues were lost. The employees were completely demoralized, and KNPC morale was literally on the floor. I remember that, although I was not in charge, I was the KNPC representative talking to the media a couple of times about the incidents. I became the visible face and the voice of KNPC for the larger Kuwait. I did it because I felt I owed it to my company. It was the start of a new phase in my career.

I placed safety as my top priority, not in words, but in actions. With Hami Hussain as Chairman, we initiated definition, uplift, implementation and enforcement of the Health, Safety and Environmental (HSE) standards in KNPC. We reshuffled our organization, two refinery managers were changed, and we enforced the application of strict rules, ensuring safety was a priority, not on paper, but in every activity of every employee and contractor of KNPC.

The Power of Direct Communication

What was the most difficult part in all this big change maneuver?
I think the most difficult part was to motivate our refineries personnel to embrace the change towards a safer operation. I applied myself to achieve that goal in every refinery.
Personally!? How?
I would establish my office in each refinery for three months, and then rotate, in order to cover all of them. I wanted to be close to the operations, and most importantly, close to the people in operations. We held daily operations meetings. That helped a lot. Our people regained motivation when they saw the commitment from the leadership towards their operational workflows. It was a true management of change, engaging all into safer operations.

Winning Hearts and Minds

Also, Maria, I think one of the key elements was, undoubtedly, the communication. We smashed barriers in communication style. Top-down, bottom-upwards, and across …we completely changed the communication style, and led by example. I would go as a peer engineer of the operators to the refinery, in my field coveralls. Every day, I encouraged managerial visibility, in daily communication workflows, until we achieved our desired target, which was to attain a world-class Safety Management System fully deployed in every refinery.

At the time, I had been DMD for 6 years, I felt I was ready for a new responsibility, but I have never asked for a promotion. The nomination and selection came on its own. My appointment as C&MD of KNPC arrived in 2004.

Top roles require to take decisions that no one else can take, only the top leaders. It can get lonely at the top. Did you feel this was true in your C&MD roles in KNPC and KOC?

It sure gets lonely at the top. What they say about the loneliness of power is true, but I applied my own way around it: I discussed challenges very openly with my subordinates, either collectively or individually. I always connect. I am not a person who makes a decision alone. Although I can lead to the decision, I want the decision to be made by consensus, with all team members having conviction. I want to gain the minds and hearts of my leading team, bringing them onboard. I would tell them *"This is what I am thinking"*, and *"what do you think?"* I was always very open.

The Driver

How would you define your leadership style?

I remember we had a consultancy firm that was looking into the personalities of the management. When I was DMD, I was assessed as an "Analytical" profile. Later on, the same exercise was re-applied, when I was KNPC C&MD, and my profile was re-assessed to be "Driver". I had clearly changed, shifting by the force of executing my evolving roles.

I knew I had evolved, and the roles made me change along my career.

Sami, I first met you as KOC's Chairman and Managing Director of KOC. How did you shifted from the downstream to the upstream sector?

The shift started with a seemingly innocuous inquiry from Saad Al-Shuaib, KPC CEO, which later became an appointment. He asked me what I thought about moving to upstream, to lead KOC. I had been already 29 years in the downstream sector, and I answered that to move to KOC was not my preference, but I also expressed to him my willingness to serve where it would be considered best. The truth is that I had some concern about how to deal with KOC's corporate culture, which is very strong. I knew KOC leadership would not easily accept leaders from outside. However, I need to face these challenges. We cannot shy away from challenges.

I was informed by Al-Shuaib that I was appointed KOC Chairman and Managing Director, on November 2007.

When I was appointed as KOC C&MD, I made a plan, focused on three objectives, and my first priority was to reduce the gas flaring. You may remember, Maria, how much we were suffering from the gas flaring in Kuwait! You could smell the gas flaring kilometers away from the oil fields. The other two priorities in my plan were the boost the availability of free gas, and controlling the costs. This is what I primarily wanted to do in KOC.

How did you steered KOC towards your vision?

To be honest, I was very careful, keeping in mind I did not want to trigger resistance. I found ways to gain the support of the people, and I did that by working closely with the management team. I sincerely valued their opinions and views. I had an excellent team.

Winning Hearts and Minds

What did you find different in the upstream company, KOC, in relation to the sector you knew so well, the downstream?

One of the differentiators of KOC, which other companies in the KPC holdings do not have, is a distinctive pride. I have not seen that pride anywhere else in our corporation. They always want the best and want to be the best. Another differentiator is that KOC truly cares about its employees in different ways than those of KNPC and KPC. In KOC, there is a clear interest in not to override the rights of the employee. I admired this very much, and I incorporated that into my own priorities.

In KOC, as in any other company, you may obtain hard work and dedication from anyone, only if you win their minds. If the executives, managers, or employees are truly convinced, is it possible to get their dedication. During the years of my term as C&MD, 2007–2013, we developed a fantastic teamwork at the executive, managerial and employee level.

Was there any element you missed from KNPC in KOC? I am curious, as very few people have the opportunity to shift from the downstream to the upstream. Sometimes, I think these are worlds apart in the oil sector.

They are not much apart. I experienced the same drivers were present in each, as well as many of the same challenges. I had many wins during my tenure in KOC, and embraced KOC in a way that made me not to miss KNPC. I only missed one element from KNPC. Although I tried hard, I could not match in KOC the focus on HSE that we had in KNPC. HSE was and is certainly a priority in KOC, but not at the level it has in KNPC. I missed that.

What main difficulties did you face as C&MD of KOC?

One of the ever-present challenges I handled was the KPC-KOC relationship. Coming from a Planning background, I knew that what planning departments are after, and how to challenge them. Convincing both sides, KPC and KOC to find common grounds for their targets and motivators, as well as the limiting factors, was at times very difficult. It was a communication journey, were my aim was to convince both entities to work towards the same goals, boosting their acceptance and motivation.

That was an ongoing and evolving situation that will always be there, at both the big or minor scale. But I had three major challenges during my tenure in KOC.

Three Challenges that Forged an Unbeatable Resilience

The most relevant one was a major well control incident in February 2012, in Raudhatain field, in KOC's North Kuwait asset. It started as a well control incident with gas leak and sour gas release (H2S gas), which spread and reached Kuwait City. We had to ignite it to control the H2S, causing a big explosion and a fire on the rig. Then we had to deal with it as a well control incident with fire. It took us a long time to extinguish the well.

Additionally, as a preventive measure, we closed a gathering center that was too close to the issue, affecting an important production volume for three months. This generated a huge concern at a national scale. The criticism was vitriolic. The public opinion, thorough the press and other media, spread false rumors that I was resigning. I faced the national and international media, in what was one of my most difficult times, solved with the support and strong endorsement from KPC and KOC as a united leadership block.

The other challenge we collectively faced was a gas leak problem in Ahmadi Town, affecting the employees of KOC who live there, near our operations. There are issues that hit you very badly, and this was one of those cases, when I felt a heavy pressure of the Parliament demanding my resignation. Again, with a focused attention to test a variety of technical solutions, we solved the issue and the pressure storm specifically addressed on me disappeared.

The third major challenge emerged during the preparations for the 75th anniversary celebration of KOC. A giant tent, one of the biggest in the world, was rented for the event. It had the capacity to welcome more than a thousand people, It was a pressurized tent, auto-inflatable, with a very modern bubble design. It was like a beautiful giant globe. But just three days before the event, scheduled for December the 22nd of 2009, a severe wind caused a contact of a tall light pole with the surface of the tent, resulting in a total damage of the tent. We were discouraged by this major setback, as the Emir of Kuwait was planning to attend. We ended up using a smaller tent, flown to Kuwait within an extremely scarce timeframe.

And what were your main achievements in KOC?

Among many achievements, such as the production of non-associated gas at about 150 million standard cubic feet (MMSCF) and the reduction of gas flaring at the fields from 11 to 1.5%, I think the top one was to increase production to three millions of barrels per day, tested in 2010. It was a very good year for KOC in every sense.

The Giants, the Young, the Challenges

Who impressed you the most so far in your career?

I have met many brilliant people in the oil and gas industry. I would like to cite Khalid Bo Hamra, Hani Hussain, Nader Sultan and Abdul Malek Al-Gharabally. Each one of them impressed me in superlative manner. They are giant people.

What advice would you provide to the young professionals in O&G?

I would advise them to first to obtain the knowledge and the experience; no one can take this away from you. First, I would tell them, focus early in your career on learning, rather than focusing on promotions. Promotions eventually will come, as right things will prevail. Even if there are shortfalls in the promotion system, eventually, the deserving individuals will be promoted. It may not be as fast as expected, but the right thing will prevail.

What will the challenges of the energy industry be in the future?
There will be environmental challenges. This is pressure about the environment, and it is already the biggest challenge to the oil industry. And I think we need to lessen the effects of the oil and gas industry over the environment, and also learn how to use the oil smartly.

What are you planning to focus on in your new role, as a member of KPC Board of Directors?
I will be focusing on leadership development, the development of a workflow or system that detect talent at an early stage—which is a major challenge—and to provide financial and commercial awareness to the technical people, as an enabler for bettering their careers.

A Shared Selfie

- **Your favorite word:** Responsibility.
- **An important historical figure:** Prophet Mohammed.
- **Your favorite theme:** There is an expression in Arabic, "laisse jarl assai", "only the right will prevail". I find myself using it over and over again.
- **A city:** I like quiet places, I am not into the metropolis. I am not even a fan of cities. I prefer the wilderness. I enjoy The Maldives.
- **Your favorite food:** My mother's Chicken in a pot. It's spicy.
- **Your favorite music:** I enjoy Kuwaiti music, but also international. I stayed with the music of the 70s.
- **Your favorite color:** Blue.
- **A personal happy moment:** When I received the news I was selected for the scholarship to study engineering abroad.
- **Who supported you the most in your life:** My wife, Sabah.

Post-scriptum

Sami Al-Rushaid will keep contributing to Kuwait in many ways. His role of member of the Board of Directors of Kuwait Petroleum Corporation (KPC), will be but one of those ways of keep his leadership flowing towards advancing Kuwait's oil sector to new heights.

Maha Mulla Hussain

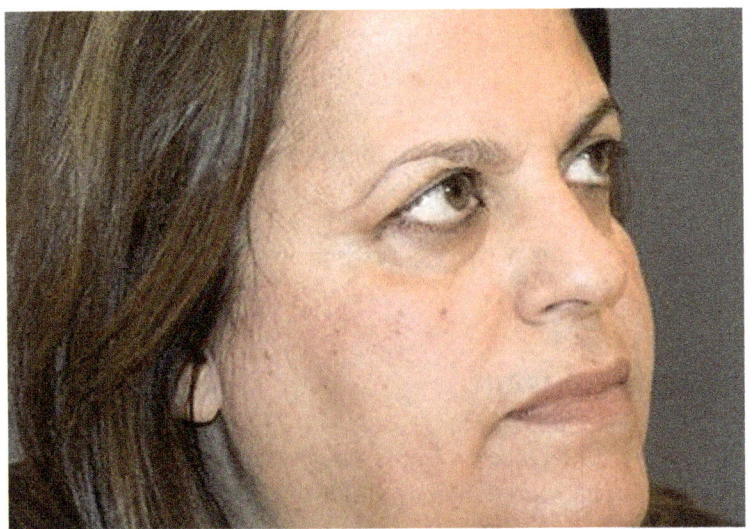

"I am committed to ethics and fairness".

A Glimpse

Women are scarce in the oil and gas industry, but even more in the downstream sector, the one related to refinery and distribution and marketing of the products. But in Kuwait, one woman, Maha Mulla Hussain, has reached the highest ranked role of the entire petrochemical industry, as Chairperson and Managing Director of PIC, the Petrochemical Industries Company of Kuwait. This fact catapulted her to be top role model in the downstream sector of the oil and gas industry in the Middle East, and in many ways, in the world.

In a remarkable career, not free of polemic challenges, she led several of the most strategic initiatives in the petrochemical and refinery governmental and private sectors of Kuwait: In addition to being the Chair and Managing

Director of PIC, Maha was also a member of the board of Equate and the board for the Joint Venture with Dow Chemical of MEGlobal.

- Bachelor in Chemical Engineering, University of California at Santa Barbara, California, (USA), 1976.
- She joins the Petrochemical Industry Companies of Kuwait (PIC), the petrochemical branch of Kuwait Petroleum Corporation, the State of Kuwait's holding, as a Process Engineer in 1976.
- Manager Corporate Planning, PIC, 1986.
- 1992, Appointed Deputy Managing Director of Planning and Finance at PIC.
- 2001, appointed Deputy Managing Director, Dty CEO, for Planning and Joint Ventures Directorate at PIC.
- Chairperson and Managing Director—CEO of PIC, 2009.
- Founder and Director of GEMS Advisory Company, Kuwait, 2014.

A Personal Snapshot

In 2012, the first conference on women matters, organized by the Society of Petroleum Engineers, International Professionals in Energy (IPEC), was launched in Kuwait. It was also one of the major initiatives for the advancement of women in the Middle East, chaired by Hosnia Hashim, and coordinated by Maria Angela Capello. For that occasion, Maha Mulla Hussain was invited as a keynote speaker to represent the national Kuwait oil sector in the opening executive panel, among other executives of high global resonance.

This opportunity gave us the best glimpse on what Maha valued about leadership. We asked her to prepare a keynote address about her journey into her CEO role, and the slides she prepared became iconic during the conference and onwards. Her main recommendation was to "know the strategy of the company for which you work". This is an example of Maha's style of work: directing her teams, her groups, and her company to reach corporate strategic goals. The IPEC conference was the perfect frame for Maha to motivate the audience, in such a way, that the next day, dozens of supervisors were explaining details of the corporate strategy to suddenly interested and inquiring employees. Female employees, that is. Those who attended Maha's presentation, and many more who received by word of mouth this impactful recommendation!

We hold a great deal of admiration for Maha, as she is grounded not only in her accomplishments and leadership style, but also on the immense resilience she showcased with the great challenges she had to face in her career. For example, when she had to face an extraordinary ordeal related to the questioning of a main agreement with Dow Chemicals, that went to a lengthy legal litigation. What would have seemed unsurmountable to many, was handled with grace and an extraordinary resilience by Maha. Resilience is one of the pillars of leadership of our book, and Maha is one of the best examples we know who exemplifies it.

If the professional career was a river, where you would come across turns, meanders, rapids, and slow waters, Maha's would have been one that traversed a high speed torrent. She advanced her professional career journey navigating her many challenges with mastery.

Arranging the Interview

Maha has retired from the government oil sector of Kuwait, and is now one founder partner of the consulting firm Gems Advisory Group Kuwait, a company aimed to provide managerial and technical support for the up- and downstream sectors, focused initially on the Middle East. In this new journey she has initiated with great success, Maha is as busy as ever, and the interview was arranged to take place at her office, in downtown Kuwait City.

A fantastic view of the many new skyscrapers of the vibrant and growing downtown of Kuwait served as the perfect framework for our conversation. From her office, we can see a glimpse of the building of Kuwait's Parliament, a body of great importance for the democracy of this country, one of the longest in the region, installed in 1962. Maha's professional destiny was directly tied to the discussions occurring within this political body, which affected decisions taken in relation to the expansion of the petrochemical industries of the country.

Our interview was focused on the personal approach that Maha took to advance her career, to highlight her leadership qualities. It was a very amenable encounter, where her candid approach to the answers ruled the conversation. We rejoiced to find a positive and progressive Maha, looking with optimism into the future of her country and the world of oil and gas.

A Resilient Leader

From the US to Kuwait

Tell us about your career, Maha, we understand you were one of the pioneer women who studied abroad sponsored by the Kuwaiti government.

Yes, I graduated as a chemical engineer from the University of California at Santa Barbara, in the USA, back in 1976. I went without my family to USA, with a couple of other female Kuwaiti students. We were the 2nd official scholarship from the government of Kuwait, as it was not common to send women to USA for study. I was part of these first two batches. In total, we were nine women across the United States. One of this group was with me in California.

Was it easy for you? How did your family cope with the detachment?

I had two options for my studies abroad, as I had to decide if I would go to Egypt or to the USA. I have to admit, it was little difficult with my father, as he was hesitant and even reluctant to let me go. But I have my brother to thank. He slowly but firmly convinced our father to let me go, because in our culture, I had to have the permission of my father to be able to study abroad.

How was your adaptation to the culture of the new setting, in the US?

I was happy from the moment the decision was made! At the time, there was nothing else I would have wanted. I desired to earn my bachelor in the States, and my preferred field of studies was Chemical Engineering.

I must say it was not easy, but I was so motivated that I merged into the group as just another student, perhaps with an accent in my English, that slowly vanished away. I was welcomed by all, students and faculty, and that made my student years easy ones. I returned to Kuwait hopeful to help my country in the oil sector.

The Training Years

Where did you started your work in petrochemicals?

After graduation, and upon my return in Kuwait, I was immediately hired by PIC, Petrochemical Industries Company. At the time, PIC was not as complex and large as today. There was the Shuaiba Fertilizer plant, and the Salt and Chlorine plant in Shuwaikh, near Kuwait City.

I was assigned to the fertilizers plant in Shuaiba, for my new-hire training period, for seven months. Afterwards, I joined the chlorine plant for the same, training. I continued there, and after progressed in my career and incremented my accountabilities at work, until I was appointed supervisor of a large group, in a role that was called Superintendent, somehow equivalent to today's Team Leader role. I was supervising the whole plant as Superintendent in 1986.

Suddenly, the Invasion

In 1990 the troops of Saddam Hussain invaded Kuwait, and we had to safely shut down the petrochemicals plants. Tanks were controlling the installations of the petrochemical industry, banning the entrance, and we could not go to work anymore. I stayed in Kuwait throughout the invasion period.

After the liberation of Kuwait, we went back to work.

Just like that?

Yes, there was work to do! All the oil industry workers returned to their roles. Our industry had to be re-built. Most importantly, our oil wells were on fire, and they had to be extinguished and brought back to production.

In what shape did you find the plants?

Without oil, there is no petrochemical industry, and to extinguish the oil wells took some time, as you know, with support from all over the world. In the petrochemical sector, we started by assessing the damages in our plants, and then rebuilding and fixing what had to be repaired. It was quite a complex task. We went on, and rebuilt the organization and the operations of PIC.

Focus on the Strategy

Then, for environmental and other reasons, the leadership of PIC decided to close down the plant near Kuwait City, so I was shifted to other tasks related to planning, and started to work at the head office in the group of corporate planning. It was one of the best experiences of my life!

When did you shift to strategic planning?

This before the invasion. It was in 1988.

I make note once more, actually, it would be impossible not to note that all leaders in the oil sector of Kuwait refer to the major milestones of their career in big chunks of time, they refer as "before" or "after". These labels are referred of course, to the major catastrophic event in the country, the invasion of Sadam Hussain's troops to Kuwait in 1990. So, all is referred to before or after the invasion.

What Is Our Vision?

At that time, the focus for the petrochemical industry in Kuwait was always placed on fertilizers, and hence the work of the strategic planning department of PIC was on expanding and maintaining the workflows pertinent to fertilizers. After the invasion, all companies were busy, and it took a while to recover the interest in the petrochemical sector. But in 1995, I was in the corporate planning, and the main question I pushed for was "*What is next*"? "*What is our vision of future*"? And with one of the most interesting strategic outlooks ever steered in the downstream sector of Kuwait, we developed the long-term strategic vision for our petrochemical sector.

Before the invasion, KPC had directed PIC to arrange for the first petrochemical planning, but 2–3 years were completely lost in arranging and fixing the great damages caused to our plants and offices by the invasion.

Then, we were thinking and focusing in the long-term and sustainability of the petrochemical industries in Kuwait, one of the pillars of our downstream sector.

Owning Every Word

How was this large-scale planning for the petrochemical sector initiated?

When I was in the Corporate Planning, I started developing the long-term strategy, with the CEO, at the time called Chairman and managing Director. I believe we did a wonderful job to set the vision for PIC, including the planning of the future of petrochemicals industries in Kuwait. This is maybe the experience which I liked the most in my career.

I owned every word of the new strategy. Nothing in the strategy was unknown to me, as I knew every word. What were the drivers, what we wanted from every sentence, and how we envisioned the implementations needed to reach our vision. That is one thing that marked my strength: I always go back to the strategy, I knew what the foundations of the strategy were, what was in it, and then, I implemented ways to develop it.

Was this a usual process in your company?

No. We were doing this for the first time. People were not even used to establishing the values, vision, the long-term strategic goals and targets. But it was needed as we envisioned our collective growth, just as a nation needs a vision. This was definitively a new exercise, which enable us to move the petrochemical industry into a new era.

We engaged in many workshops to streamline what was to become a common vision. It was not easy. There was a lot of reflection, persuasion and convincing. You may imagine there was a great deal of discussion and debate. And when we finally reached an agreed vision for the future of the petrochemical sector in Kuwait, there was a great deal of satisfaction, motivation and enthusiasm. It was a vision of growth, of modernization.

The Language

Then, came the need to communicate this vision to all in the company. We had a vision, but how did we communicate to all, to every employee in the company? We established a communication campaign that encompassed all groups and functions of PIC.

What do you consider was the most important factor for communicating the strategy?

The language. We needed to literally translate the strategy and its impact for each group in the company to be understood in their own terms. With examples, in order to explain the content and the goals ahead.

It is a complex exercise, I can imagine the many challenges. Like storytellers, you had to adapt the message to the audience.

Yes, we had to make sure the goals and possible implementation path to achieve our targets were fully understood by all, from Operations to Human Resources, from every Plant employee to the managers. I am very satisfied with the results achieved. I would say the vision and goals, our strategy was communicated efficiently, and all in PIC were aware change was coming, to attain that vision we crafted.

And what happened after a vision was crafted?

From that strategy came another project that was important for me: PIC needed to redesign its internal processes, to be aligned with world best practices, to achieve the best efficiency.

We created the ALDANA projects, which was a "business process redesign", with a group of consultant firms. We tackled finance, operations, maintenance, HSE, Human Resources, in a nutshell, all major functions. We needed a major overhaul, an important redesign.

I led the approval processes needed to implement these changes, in the Aldana project, with our CEO in PIC, and also with Kuwait Petroleum Corporation's CEO at the time, Mr. Nader Sultan. PIC was and is a subsidiary of KPC, so all approvals for major plans and projects had to be obtained from KPC, to enable the implementation. Sultan told us "This is needed, it is a fantastic plan." We were thrilled.

Did anyone had any doubts this overhaul was needed?

At the beginning, the leadership of PIC, my colleagues Deputy Managing Directors and the CEOs questioned and challenged the need to redesign all the process, but the added value of reshaping the processes to achieve efficiency was huge, and the numbers spoke volumes. We at the Planning groups were certain and confident there was no turning back, that the only way ahead to improve was to change. We took four years to do it. I was appointed Deputy CEO—at the time, Deputy Managing Director for Planning and Finance.

The Benefit

To change the way people work is not an easy task. Especially if you are dealing with individuals who have been in their roles for 20 or 30 years, doing their work in a specific way over and over again. So, we initiated with a massive communication campaign, to explain the new way of working, but especially to highlight the benefit of the new way of working. The "*whys*", and the "*hows*".

Do you remember any example where this change resulted in a beneficial improvement?

Sure! I always highlight that Health, Safety and Environment (HSE) is the best example of this change. For the people working in HSE and for all in operations, wearing a helmet was sufficient as a protection and safety measure at work. After the application of Aldana, we completely changed the handling of HSE, and added not only complete sets of personal protection equipment (PPE), but also HSE's workflows were implemented for collective protection and monitoring. We implemented HSE audits, monitoring, site visits, standards, and quality control checks of all kinds.

We soon started to implement the enhancement and expansion strategies for PIC. I then continued to work mainly in planning and strategic initiatives, and in 2001, I was appointed Deputy Managing Director (Dty. CEO) for the Planning and Joint Ventures Directorate at PIC.

The leadership and resilience is notable in this period you are describing. And this was at a time when women did not reach important roles in the sector. Was it difficult for you, to be the only woman at the board?

I learned I could do it. I felt part of the team, so for me it was not difficult.

The Fairness Equation

And if you were to choose from your many personal strengths, which ones do you consider propelled your success?

Strong work ethics. My commitment to work ethics. And then fairness. People value fairness. The employees have expectations that generally differ a great deal from the plans the company has for them. Generally, they think they are the subject of great unfairness, that some bosses show obvious preferences. In Kuwait there are still preferences, notable in the work environment.

And yes, there may be differences in the groups, but as a supervisor, you have to be fair and opportunities need to be available for all. Decisions need to be grounded on professional criteria and not preferences ungrounded on a technical basis. People expect and deserve fairness.

I stick to fairness, making a conscious effort towards being fair. I gained the respect from people as I was fair.

This is probably what led you towards an executive role.

I am convinced it was key for that. My first executive appointment was during the execution of the ALDANA Project, it was during those years that I became Dty. Managing Director of Planning and Finance.

Did you have to learn Finance?

Yes! I was responsible for the forecasting of the company's financials. We engaged in detailed financial planning, preparing analysis of how the company would perform in financial terms, given several scenarios of production, of investments, of sales, of energy demand. How will the financial health of the company look like in the short- and mid-terms? These questions drove the activities in financial planning.

At the beginning, I was leading corporate planning and the financials. Then, I headed Joint Ventures, the JVs, as well. We reached seven JVs, in this growing expansion. I was assigned as Director of Planning and JVs, handling these functions, and IT, and other connected activities across PIC.

What were you particularly proud of during those years?

I was proud of the growth of PIC. Also, of the implementation of the PIC strategy, which was to grow in petrochemicals inside and outside of Kuwait by building new plants or acquiring existing plants, with a strategic partner. This was always the focus.

In the narrative of her journey towards the highest roles at PIC, we cannot but observe that Maha Hussain embodies the main characteristics of leadership we consistently find in top executives: the need and ability to communicate a vision, the focus on strategic goals, a fearless attitude to change the status quo, and handling of the company as it was their own. But besides leadership, leaders also embody resilience. And I am about to ask a question to Maha that will trigger a reflection in me about resilience. A great deal of resilience.

Dow

Maha, I want to hear directly from you what happened with the partnership with Dow, as I know this was a complex matter for Kuwait Government, which occurred during your CEO tenure.

When I was Deputy Managing Director Planning and JVs, we launched a strategic alliance with Dow Chemicals, aimed to implement our vision, to expand our petrochemical industry, with strong cooperation of key partnerships.

PIC had the idea of acquiring 50% of existing assets that could be available and would match our business objectives in the United States, in Europe, in South America and in Asia. We studied this opportunity in detail, and came forward with a proposal, that we submitted for approval. We distributed and presented it to all the approving instances, namely, internally in PIC, as well as in KPC. Finally, we submitted it to the approval of the Kuwait's Supreme Petroleum Council (SPC). We submitted the preliminary study, informing the results of our analysis, and our proposed way forward.

We announced a Memorandum of Understanding (MOU) with DOW in 2007. The MOU stipulated that we would acquire 50% of DOW's polyethylene and other assets, in what was to be a negotiation of billions of dollars. We did not receive any negative reaction in Kuwait.

During this time, I became Chairperson and Managing Director of PIC, and continued to advance this main strategic goal. We proceeded to the second stage, and I led all efforts to lead the due diligence, taking approvals from all pertinent boards of directors: our own in PIC, in KPC, and then in the SPC.

After we took approvals, we announced that we would execute the agreement with Dow, and the press and political forces of Kuwait voiced out this agreement was a surprise, that it did not seem properly handled. They expressed vividly that in their opinion some elements had not been disclosed, raising severe doubts and creating an extremely negative image for this alliance.

Although we followed all necessary approval processes, the political opposition to the initiative was so strong, coupled with the economic situation, that the higher Ministers Council directed the project had to be cancelled. They issued a resolution to drop the project. It was a huge set back.

The Dow acquisition would have brought many opportunities for the country. In particular, for the growth of our petrochemical industry, our employees, who could have worked with new technologies in and outside Kuwait. Dow would have established an R&D center in Kuwait, and there were many other benefits.

The resistance was mainly led by politicians. Dow was not pleased that the project was cancelled, and they proceeded to litigation. And they won. The downside of this is not only losing the opportunity to grow, but the penalty fees we had to pay to Dow, on the order of billions of dollars. It was a huge setback for the nation.

Newspapers started attacking some leaders in the O&G sector, criticizing them. The articles highlighted the damage of losing the litigation, and that the leaders who led the Dow deal were still in their roles. The minister of Council issued a resolution, that leaders involved in Dow had to be suspended from work. Then, the Minister of Oil of Kuwait had to deal with this big issue.

What was the Minister's decision?

When Mr. Hani told me about the decision, he presented me with two options: to be retired or suspended.

"I Will Not Resign"

The decision was made in May, and my term was going to finish in September, as I had reached retirement age. But I decided I was not going to resign. I had not done anything wrong.

You made the decision of not resigning.

Yes, and I informed the Minister, who then made the decision of retiring all executives who had completed 35 years of service, and I was one of that group. Those were truly hectic years. Lots of pressure.

The oil professionals at all levels were convinced, as we were, that we did the right thing, following our procedures and approval channels. But there was negative press coverage, and the opinion matrix of the general population in Kuwait changed. The citizens were gossiping that this was a mistake made by the professionals. They said that we had made these decisions in the closed diwaniya environment.

Why do you think this reverse in the opinion matrix towards a negative perception was feasible?

I have come to understand that we, the professionals and leaders of the oil industry, maintain our communication within the confines of our corporations. But politicians talk very loudly. They express their opinions in a wide scale and volume. They have echo. And this in true in Kuwait, not only for Dow, but for everything.

Politicians have hit others in Kuwait. Innocent people.

I did not feel worried for me. I knew I executed my role with due diligence, but I felt truly sad for the country.

Was your family worried about you?

Yes, they knew, and they were worried about me. My health, my capacity to take on all the attacks towards me. They worried about the consequences of those politicians for my persona. That they could go further and press charges, be judged.

I have not finished yet. Prosecutors are now compiling information, 10 years after the facts.

It must be very difficult to cope with all this situation. I cannot imagine how you coped with all these difficulties. You are strong.

I am not strong with things that hit my country, I become emotional. But at work, and doing the right thing, yes, I am strong. Especially if you have a good team around you, like I had.

It was good to work and lead at PIC, because it was a team effort. You will find that in some companies, the top leaders do not shape a team. But in PIC, we were truly one team. I am proud of that. My strength came from this.

Those Who Disappeared

What is the strength of your network now? Do people contact you as often as before?

After retirement, some people keep contacting you. Asking about you with a genuine interest. But others disappeared. Literally! There were people who I thought were genuinely interested in me, who greeted me during Eids and festivities, from whom I heard nothing ever again. That's life.

So, you learn to differentiate between those liaisons who were contacting you just by interest, because you had a relevant role, from those who do care, the loyal ones. I guess the Dow and the retirement years were a filter that helped me value and recognize true loyalty.

The Consulting Firm

When we retired, from the beginning, I discussed with two of my ex-colleagues, CEOs of other companies in the Kuwait oil sector, to launch a consulting firm to remain active in our profession and keep contributing to Kuwait's future from the private sector.

We came to an agreement with only one sentences or two sentences, because we knew each other so well!. After enjoying the retirement for a year, with our families, grandchildren, and travelling, we launched "Gems". We wanted to get in touch with the industry, as we hold knowledge and experience from upstream to downstream. We can help companies and international enterprise to grow.

Doing something for Kuwait is a main driver.

Maha, who supported you the most in the oil industry?

When I was at corporate Planning as a senior engineer, there was a CEO of PIC, Khaled Bohamra, who is now a member at the Kuwait Petroleum Corporation Board. I would like to think of him as a women's supporter. He is a very open-minded professional, who truly supported me at work, as well as other women. He was keen in pushing for women into leadership, to decision making, giving challenges to them. I still believe he made a shift in me towards finding within me the leadership I had.

He empowered and encouraged me. "You will do the presentation to the Board, Maha", he would say, and I was just a senior planner. Sometimes we need this kind of encouragement.

What are the challenges the energy industry will face in the future?

We saw what happen with the shale gas and shale oil. All predictions about how much gas would come into the market were shaken and changed by the new technologies to produce from shale. What would be the renewable energy contributions into the future? Those are still unforeseen.

The challenge is how to predict the new technologies that will make renewable energy cheaper and available for the people, as this will revolutionize our sector.

What message would you like to send to the young generation?

If you are in an organization you need to understand the company strategy, what is expected from you, at your current position to contribute to that strategy. Commit to hard work.

I always went to my work happy. I loved my work, so I would recommend to the young generations that if they do not feel happy to solve any issues. Do not go to that work that does not make you happy.

A Shared Selfie

- **Your favorite word:** Fairness.
- **A city:** Besides Kuwait, London.
- **An important person of your preference:** The listeners. I like the people who listen rather than those who talk.
- **Your favorite food:** The Kuwaiti "Majbous".
- **Your favorite color:** Blue, but I change a lot about colors, and do not have a particular favorite one.
- **A favorite song:** I like the old ones. Oum Koulthoum is still my favorite singer.
- **A landscape:** Italy. Tuscany! The grass, the hills, it is amazing.
- **A personal happy moment:** Any moment with my grandchildren. I have three: two girls, Maha like me, and Tala; and a boy, Mejren, who I call Mijo.
- **Who supported you the most:** My husband, Hamed.

Dr. Kamel Ben Naceur

"Projecting into the Future".

A Glimpse

"*As a result of major transformations in the global energy framework, that will take place in the next decades, renewables and natural gas will be the winners in the race to meet energy demand growth until 2040*". This is an assessment contained in the November 2016 edition of the World Energy Outlook, the International Energy Agency (IEA)'s flagship publication.

The IEA is an autonomous organization of 29 countries, aimed to ensure reliable, affordable and clean energy. In 2015, IEA sought an executive with extensive experience in technology and clean energy for the Role of Director of Sustainable Energy Policy and Technology. It is no wonder that they found a perfect match in Dr. Kamel Ben Naceur for such a strategic role. He is now leading IEA efforts to modernize the approach to energy technologies. He guides the organization's outlooks, providing a cornerstone reference for the energy world.

Kamel Ben Naceur, is one of those leaders whose career was shaped by an extensive international exposure in energy, as a former Tunisia Minister of Industry, Energy and Mines, previously a Schlumberger executive, and now as a global visionary at IEA developing possible roadmaps that will take us to the future using the energy resources of the planet well.

Dr. Ben Naceur, has more than 35 years of experience in the energy private and governmental sectors. He is one of the few individuals in the world to hold degrees from two of France's top educational institutions, the École Polytechnique and the École Normale Supérieure. After he stepped down from his ministerial role, he received the highest distinction of Tunisia's government, the National Order of the country. Kamel served on several boards of international businesses and organisations, and has received numerous medals and recognitions. He is the co-author of 13 books and more than 120 articles of technical and managerial nature. He is a colossal role model for the young generation working in energy.

- Is a national of France and Tunisia.
- 1976 studied at the École Polytechnique and the École Normale Supérieure, in Paris.
- 1979, enrolled in the École des Mines in Paris as a research engineer.
- December 1980, joined Schlumberger.
- 2004, initiated the Schlumberger business unit to develop CCS (Carbon Capture and Storage).
- 2009, Chief Economist of Schlumberger.
- 2011, President of Schlumberger's Technology operation in Rio de Janeiro.
- 2014 Tunisia Minister of Industry, Energy and Mines.
- 2015, Senior Advisor and Vice-President for technology of Schlumberger, based in Paris.
- 2015, IEA International Energy Agency Director of Sustainable Energy Policy and Technology.

A Personal Snapshot

Dr. Ben Naceur is one colossal leader, although his demeanor is down-to-earth, open and cordial, humble. He makes you feel at ease, as if you would have known him for a long time, including you as a natural part of the conversation.

We have known Kamel for decades, during his years in Venezuela, and then later in the Middle East. Besides business connections, we know him most particularly for his deep involvement with SPE, where he has volunteered extensively, serving at the Board level, and in so many other leadership roles.

We are proud to have Kamel Ben Naceur among our most esteemed liaisons, and the conversation with him enabled us to dig deeper in the history of Kamel as a leader. He is resilient, adaptive and successful. He is both innovative and focused on the people.

Arranging the Interview

The response from Kamel to our request of interviewing him came as a direct, swift and amenable yes. We were delighted, considering his hectic agenda of traveling and global meetings required in his role as Director of the IEA, in Paris.

We set a meeting and Kamel was, as usual, relaxed and willing to share his point of view. Once more, we were learning from him.

One of Those Things

Tell us about your career, Kamel.

I was born in the city of Gafsa (Tunisia), raised in Tunis, and I completed high school in Tunisia. I had the privilege of being selected among a reduced number of students to receive a scholarship from the government, as I was in the top students of my cohort. The Tunisia's government had a system in place to send the young students to programs in France, Germany and other countries. I went to France, to the first-ranked preparatory school (Lycee Louis le Grand, Paris) and also was very fortunate to receive the entrance to the all of the top Grandes Ecoles. It was the first time this happened in Tunisia. I had in my cohort of Lycee Louis le Grand and Ecole Normale

Superieure two students who subsequently received the Fields Medal (the highest honor in Mathematics).

Were you in your early twenties?

I was the youngest of my group, at 17 years of age. We were 20 students from Tunisia sent to France at the time to compete for the top Grandes Ecoles.

Cold Winter in Paris

I am surprised about how young you moved to study in France. What effect had this change of environment in you?

I had two older brothers, who have travelled to study in France before me, in the same system. It was an extraordinarily good introduction for me, as I knew what to expect. It was an outstanding experience in every sense. When I sent there, in 1973, it was the time of the first big shock of oil prices. All countries were affected in one way or another. In France, we were tied to the student funds provided to us by Tunisia, which were curtailed.

Additionally, France was coping with the rise in the price of oil. In the first winter of my life in France, it was the coldest one, as the heating was severely reduced! I would say this develops your resilience, because you have to be extremely resilient to face not only the cultural change, but also to face physical discomfort to move on, motivated only by the willingness to achieve your degree.

The Thrill of Multiculturalism

Immediately after my graduation, I was hired by Schlumberger. I started to work in the research and development center, near Lyon, at St-Étienne, in France. It was an extraordinary experience, providing the opportunity to literally build my multinational skills, as the teams were truly integrated with professionals from several regions of the world.

I worked four years there as a researcher. I was Project Manager and then Section Manager for Subsurface Studies, mostly focusing on well stimulation and cementing.

This is one of the main technical services of Schlumberger, if I may say!

Indeed. It was a central department at Schlumberger. We were very pleased, as the active research that we were leading resulted in publishing numerous technical papers in peer-reviewed conferences and journals. Some of the publications of this period made it to the prestigious magazine,

"Nature," which was a tremendous success. We aimed to implement only those successful developments that looked promising, and we had many!

Showcasing R&D Round the Globe

The role I had, coupled with the frenzy of activity of the researchers in my team, provided me with the magnificent opportunity to showcase Research and Development new products and ideas around the world.

In 1985, I was sent to the Schlumberger center in Tulsa, Oklahoma, to initiate an exchange of people, and engage directly in knowledge transfer. This was very interesting, especially in 1985! This was not common at all, but we understood the key was in sharing the learnings and expanding the knowledge.

How was this vision crafted?

I would have to go back to 1981, when I started working in the international network of the oil and gas industry. I remember clearly, because it was in 1981 that I co-authored my first technical paper for the ATCE, this is, the Annual Technical Conference and Exhibition of the SPE, the Society of Petroleum Engineers. It was about the implementation of new technology for fracturing in Peru, in the Talara province, in the north of the country. Then, I worked on some key projects the company had in Libya and Italy. I gradually started to understand the importance of sharing the knowledge globally and learning from colleagues. Each one of us was working in different settings with regards to the technical characteristics, legislation, and clients' appetite for technology. And if we could build from this great strength, to benefit all and accelerate implementation, it would push our knowledge even further.

The Leadership Call

In those years, I was getting immersed in a new era, shifting from individual projects, of 10–20 thousand dollars in the public research, to multimillion dollar projects with global outreach. What we applied in Libya could be key in Brazil, and vice versa.

It took me a bit of time to adapt, but within two years, I was promoted to a managerial role, and I progressed from being a project manager to becoming Section Head. I would say this is the moment and exposure that propelled my leadership path. I had ten direct reports in my Section.

Anticipating the Future

What were the primordial elements in your leadership?
(Kamel paused here, and very systematically, listed for us the three elements of leadership, as if he were to prepare a report or a presentation. It was great to see the IEA Director analyzing his own story in real time for us!)

Firstly, to communicate with my scientists. It was more than a one-way communication on my end. The key was to understand their highly technical explanations, translating them into action plans and possible projects. I realize now that I became an efficient translator.

Secondly, to encourage those researchers interact among themselves fruitfully. Even if I was not there, even if they were from different specialties related to an understanding of the subsurface. As geologists, sedimentologists, geophysicists, reservoir engineers, petroleum engineers, drillers, they had to interact to collectively achieve progress.

And third, the ability to project into the future.

Solving the Problems of the Future

If you are in R&D, obviously, you are not solving the problems of yesterday or even today, you are trying to anticipate how to solve the problems of the future. I recall as an example discussing with the marketing department the specifications of the next generation software for simulating cementing operations. What they had in mind as a 10-year challenge (e.g. simulating multi-fluid flow including U-tubing) was actually accomplished in less than a year!

How did you expand the role of R&D within Schlumberger?

I would like to think we expanded the role of R&D also in the workflows of our main clients, the IOCs and NOCs, International and National Oil Companies. How did I envision this expansion?

It was not an immediate expansion. I stayed in a leading role for R&D for nine years, taking care of teams in three centers: France, USA, and the United Kingdom.

Expanding the utilization and implementation of R&D results was attained by attaching R&D to operations, in different continents, working from a country-focused effort to a regional one, and finally to implementation of a global scale. We developed area marketing managers for technology. This helped to reach the objective I had in mind, of sharing the R&D developments from a country to a global scale. Again, it was important to anticipate what were the main problems to be solved in each region for the future. In 1999, I became the worldwide vice-president of marketing for Schlumberger Dowell, working with our technology centers around the world to provide the specifications of next-generation technologies, as well as our regional operations to accelerate the deployment of technologies that had just been introduced.

The Right Hand

It becomes clear that you developed very important traits of leadership in these years in technology management. At what point you start tapping into strategic roles?

The corporate vision for Schlumberger in R&D, that I helped to shape, included as early as 2002 the creation of a new business unit dedicated to Carbon Capture and Storage. In that role, I had the opportunity to expand significantly my business perspective, from an oil and gas one to an energy-wide one. It also put me in contact with new stakeholders, such as legislators, regulators, and non-governmental organization (NGOs). From that point, I moved to the position of Chief Economist, responsible for strategizing and forecasting scenarios and outcomes for all of Schlumberger.

That was a key role with a tremendous growth opportunity. I was in charge of providing input for strategic decisions. Those related to divestments and acquisitions.

You were the right hand of the CEO of Schlumberger!

Yes, I guess you may say that I was the trusted advisor of Andrew Gould in terms of market analysis and predictions. It was one of the most interesting and challenging roles I have ever had, especially since Mr. Gould's communiques during the quarterly earnings reports were closely watched by the whole investment community.

The Honorable Minister of Industry, Energy and Mines of Tunisia

Kamel, let us shift gears, and learn about your experience as Minister of Tunisia. How did you become aware you were considered to be a minister of the country where you were born?

It was a phone call.

That call came on a Sunday morning. I was doing my morning jogging. I do not know why I had my mobile on me, as I usually leave it at home, in order to enjoy sports time fully.

I received the call from a person who introduced himself, and I knew him by name. He said "I am the newly appointed Prime Minister of the new government of Tunisia, and I am considering you for a position in the cabinet". I immediately asked how much time I had to think about it.

The answer was very simple and direct: "You have until the end of the day".

Obviously I had to consult with my most important stakeholders!

We laughed, and I anticipated who these two more important persons were. And I was right, when he continued, with…

My wife and my boss.

My decision was of course very much dependent on their feedback and vision about this.

My wife was extremely supportive, and said: "Do what you feel, but know we will support you", encouraging me to accept. As a Tunisian herself, she knew our country needed to re-establish the internal processes in every sector.

Then, my boss at Schlumberger was also very supportive, for which I will always be grateful. He said that they understood this Ministerial role was somehow uncertain, and that they would let me take it, maintaining my position in Schlumberger for me, to be discussed upon my return to the company. They were really flexible, open, and understood my drivers to serve my country with my best effort. I felt I was needed, and was grateful that my company of so many years enabled my contributions to be initiated in a direct way to sustain Tunisian re-birth as a democratic country.

These two answers were encouraging, making it pleasant for me. Then, the most difficult thing was to announce it to the team. So, two days later, I was on a plane en route to Tunisia to get to know in person all my colleagues at the Cabinet and to participate in a formal interview conducted by the Prime Minister.

The preliminary meetings went very well. What I received were two positions merged in one: I was appointed Minister of Industry, Energy and Mines of Tunisia.

The Way Ahead for Tunisia

How did you get started?

Picture that I had left Tunisia 40 years earlier, only returning back for family reasons. Before this change towards democracy, I had no interest in working in Tunisia, but now I was pleased to have been given the great opportunity of helping my country and to give back.

I was very privileged to initiate my role as Minister in the first two weeks of the new government. I worked with a very small team, as we were only 4 to 5 people, in charge of setting the priorities of the government, for the inaugural speech of the new President. That was, for me, the first way to contribute to the political program. This first task obviously enabled me to directly and swiftly catch up with what was happening in my country.

Almost immediately, I had what was to become my first truly political experience: To get the government approval, by the National Assembly, of the plan we recommended. This was very special, because the members of the National Assembly questioned everything the government tried to advance, on the grounds of unsuitability. After 16 h of debate sessions, the plan was accepted with a strong majority, and we could go on to work on our individual responsibilities.

Giving Back

One way to describe my most important satisfaction with my role in Tunisia, was that the role as Minister provided me with the opportunity to give back to my country! What I received from them in education, in Tunisia, and through support for my studies while I was in France, could be somehow thanked. It allowed me to show appreciation and retribution.

Was it easy for you to lead in your country's governmental institutions? How did you adapt to the new setting?

Not at all. For me, it was a complete change, from what I was used to doing. But if I had to do it all over again, I would! One thing that was very rewarding were the relationships formed at the executive level. I met extraordinary individuals who are role models in their own right and in many ways. It has been two years since I stepped down as minister, and we still liaise and meet when we can.

My tenure lasted one year and two months; it was extremely intense, with many challenges ahead of us. One of those goals was to organize the country's first fully democratic elections. We did that, and there were no objections on the process, nor were there issues from the security side. I was very proud of this achievement.

The second challenge was that it was very important to reestablish or envision economic strategies to develop the country in every sector. In the energy sector, the priority was to put together a vision for the country in term of energy priorities and rationalization of subsidies.

Then we needed to reinitiate the oil and gas exploration and production activities, and finally, pass a new law on renewable energy. We did all these during my tenure.

Arabic for Business

Kamel, what an amazing journey! And what were your personal biggest challenges for you as a leader, in such a different environment from the one in which you were so used to working?

My biggest challenge was about the way of working. During most of my career, I utilized the English language and I relied on electronic communications, like emails. For me, to shift to Arabic, a language I was not using for work for over 40 years, and to not use emails for work communications was extremely difficult. Also, it was also challenging to learn how to deal with different stakeholders in the society: the Parliament, the Unions of Workers, and the international partners for Oil and Gas, as part of the Government.

That would have seem unsurmountable to many! Would you like to share what skills or strategies did you apply to overcome these mountain-sized challenges?

Well, what I used were three main skills of leadership: Listening capability, strategic thought, and my knowledge of the business. I would like to emphasize that the strategic thought was in this case again fundamental, I am specifically referring to the capability of anticipating, of thinking ahead. I cannot overestimate the value that my experience in the private sector of oil and gas provided me with for my role of Minister.

(At this point, he laughs) Why you laugh, Kamel?

Because I also want to tell you that I played chess from a very early age, and I think this has shaped my perspective of things, providing me the ability to anticipate moves. It is kind of funny, but I just remembered that I am thinking ahead about your next question, as if your interview was composed of a succession of chess moves.

Oh! Ok, so here I go, with another move. What were the two elements that you consider were the peak moments of your ministerial role in Tunisia?

One of the moments I cherish the most was to receive the highest distinction of my country, the National Order of Tunisia. That was one of my best memories as Minister.

Another stellar moment for the future of my country, not for me, was passing the law on renewable energy.

Back to the Future: The IEA Experience

When I ended my ministerial role, I returned to Schlumberger as a Vice-President Strategic Advisor, a role custom made for me. Almost immediately, the opportunity of working at the IAE came up, as they advertised they were looking for a person who would take two of the biggest directorates in the IEA, merging the energy and technological outlooks into one accountable directorship.

I was very interested, so I discussed it with my company's management, and it was agreed that I would apply for the IEA position, as it related to the business objectives of my Strategic role. There were in excess of 150 highly-qualified individuals considered for the role, so I was very pleased to be selected.

As the Director of the IEA, I have now the responsibility for the all long-term outlooks in the agency, including the World Energy Outlook, Technology Perspectives, 2050 and beyond. We are providing different scenarios, one of them based on the new policies, which is not far from what other energy outlooks like the BP outlook is projecting, in line with the pledges of Nationally Determined Contributions (NDCs) made for the Paris Conference of the Parties (COP) 21. We already have gained credibility, and have some companies that are using our scenarios. For example, Total's CEO is using in his presentations our 2-degrees scenarios, as a reliable reference for Total's projections.

My current role also encompasses the technology coordination of 39 programs that cover all the energy systems of the world, from Oil & Gas to Fusion Power. This includes renewables, utilizing energy efficiency, and the coordination of the input from 6,000 experts around the world, who collaborate in progressing technology in those areas.

This is a three-year tenure position. I am half-way through, having completed 1.5 years as of today. One of the projects that we launched is "Women in Clean Energy", targeting an increase of the involvement of women in clean energy, through the IEA.

What would you say is your current biggest challenge at the IEA?

Coming from the business world, I was guided by quarterly and yearly reporting styles, where typically the strategy looks at the next 5 years timeframe. Now, instead, we have to project 40 years down the road, to properly advise governments, business, and others about potential outcomes for that timeframe.

You asked me about challenges, but I prefer to speak about achievements. I will tell you that so far, I think my biggest achievement was to be able to bring to the IEA a refreshed organization. It's now the clean energy ministry, an organization that constitute 24 countries in one body, that represents 85% of the investment in energy, hosted by the IEA, in terms of geographical business.

Also, it's an achievement to have updated all scenarios of IEA to include Paris COP 21. That is a major step forward for the way we look into our future in energy, and I am very proud of that.

What Would You Say Are the Main Challenges of the Energy Industry in the Future?

I will reply to you about the whole of the energy industry, not only oil and gas. I would say the main challenges are:

1. Energy access. Because today, 1.2 billion people do not have access to electricity, and almost 3 billion people lack access to modern energy sources for things as basic as cooling. That is the number one challenge.
2. The second is producing sustainable energy, with the lowest emissions.
3. The approach our industry takes on the high variability of oil prices. How can we handle the volatility cycles? Can that be managed? That can be somehow better managed. I do think that if the facts are provided to the different stakeholders, highlighting the implications of their choices, different decisions would be taken. Currently, the strong reductions in investments that always follow the low price valleys, is going to lead to continuous volatility in oil prices. We have witnessed 3 continued years of low investment, so in order to capture in the future we would need to increase the investment today. Just to give you one example, in the last three decades, the average new oil discovered per year was 15 billion barrels/year, but in the last 3-years it has gone down to 5 billion barrels per year. Will we be ready to produce when the price is more attractive? Even

if the shale oil comes back, that will only bring an additional 2 million or maximum 3 million barrels to the market. That is not enough to balance the energy demand.

What Would Your Messages Be for the Future Generations?

I would propose a few points to them:

- One is to enhance their collaboration efforts. Volunteerism, in not-profit societies like the Society of Petroleum Engineers and others, are excellent ways to developing those competencies. I encourage them to participate.
- To develop their soft skills as early as they can. Especially the ability to listen, and the ability to adapt.
- To foster the development of their ability to project into the future, in different scenarios, and with different options.
- To be more flexible, as incrementally, the conditions of countries and the market will change.

In general, I worry about our speed to be able to adapt to an accelerated and changing world, but the adaptation is natural in humans, so I remain extremely optimistic.

The error we should not make is to judge our youth using our own perspective. When I started in the professional work 35 years ago or so, the idea was to start your career with 3 to 5 companies, and contribute in a meaningful way in each company to achieve results in order to be promoted. Now it is completely different.

A Shared Selfie

- **Your favorite word:** Vision.
- **A city:** Paris. And I know in my heart and in my mind that my favorite city in the whole world is Jerusalem, a city I still need to visit, discover and feel.
- **An important person of your preference:** I have two. The first one is Antoine de Saint-Exupéry, the author of "the Little Prince". In terms of political figures, I admire very much the first president of Tunisia, Habib Bourguiba who had the vision to define a new role for women in an Arab

country, with education, and promoting the role of women as a primordial element.
- **A personal happy moment:** My marriage! We had more than a thousand people gathered for celebrating our wedding. It was a whole week of feasting, in Tunis. It was truly the most special occasion, happy moment of my life.
- **Your favorite food:** Seafood. In particular, scallops.
- **Your favorite color:** Blue.
- **Who supported you the most?** My wife. We have been through challenging times, and moved all around the world.
- **And your favorite music?** I like different styles, from the Arabic world, I like very much Asmahan and Mohammed Abdel Wahab, both traditional and old singers. From the Western culture, a musician I like is Elton John, and…I just love Latin-American music, in particular the merengue. You know? I have lived in Venezuela… you have to love salsa and merengue in Venezuela.

Post-scriptum

Saying goodbye, Kamel asks me about the situation in Venezuela, recalling when he lived there. He mentions to me that he was always surprised by the capacity of Venezuelans to handle the short and long term perspectives with equal facility. They could plant Uverito in a man-made industrial plantation, that required more than 30 years to become a source of timber, but they could also just think and care about the three main priorities in life.

I asked him, with some hesitation…."which ones?!"

And he continues in Spanish: "Desayuno, almuerzo y cena!" (Breakfast, Lunch and Dinner!)

We laughed in loudly and friendly way.

What a fantastic way to end our meeting, finding a common ground, in this case, Venezuela. Kamel has a strong reputation also for his people skills, not only for his insightful capacity to project into the future. No doubt!

Sheikh Faisal bin Fahad Al-Thani

"I believe success is to never stop learning".

A Glimpse

Sheikh, also spelled Sheik or Shaikh, is an Arabic title of respect dating from pre-Islamic history. It is currently used for the royal families of the Middle East, as is the case of the leader we are about to know better in this chapter.

Sheikh Faisal Al-Thani is from Qatar, one of the giant oil and gas exporting countries of the Arabic Gulf.

He is the Deputy Managing Director of Maersk Oil Qatar, which is the operator of the largest oilfield, A Shaheen, contributing to 40% of Qatar's oil production.

He has accumulated more than thirty years of experience in the oil and gas industry, in a variety of high ranking roles. His achievements catapult him to prominence as a role model for resilience and leadership well beyond his country's geographical frontiers. From reservoir studies to planning; from Qatar Petroleum to Arco; and from Anadarko to Maersk, Sheikh Faisal has been leading organizations in the oil sector towards success.

Moreover, his career path includes an astonishing number of academic achievements. He has a Master of Science from Bath University, and is a PhD from Leeds. He has been invited lecturer for the Fulbright Program, the top international educational exchange program sponsored by the U.S. government, lecturing at the University of Colorado at Boulder. He is an author of numerous books, in a topic in which he is recognized as a world expert: risk management. His books are used as subject texts at universities like Manchester and New York University. Some of his books have been translated into Chinese, and are currently used at several universities in China.

- Graduated as a Petroleum Engineer in 1986.
- Joins Qatar Petroleum (QP) after graduation, assigned to offshore operations March 1987.
- Joins QP Reservoir Studies, in charge of reservoir simulation, 1989.
- 1992, Contract and Planning department, QP.
- Arco Qatar, Deputy Managing Director, 1998–2000.
- Master of Science from Bath University, 1998.
- 2000–2001 Operations Manager BP, South North Sea.
- PhD from Leeds, 2003.
- Fulbright Fellowship 2003–2004, Colorado School of Mines.
- Anadarko Qatar, 2004–2008.
- Maersk Oil Qatar, Deputy Managing Director 2008–2017.
- Author and co-author of five books on Risk Management.

A Personal Snapshot

Hosnia met Sheikh Faisal as one of her peer Executive Committee members, during the preparation stages of a large conference in the Middle East, organized by the Society of Petroleum Engineers (SPE). Representing the oil and gas sector of Qatar, Sheikh Faisal has always been very active in promoting the exchange among countries of the Arabic Gulf, and was keen in ensuring the highest technical quality for the conference that was taking shape at the time.

What was a casual professional liaison grew into an active ongoing professional exchange, promoted by membership in several international conferences executive committees.

It is a pleasure to include Sheikh Faisal in our book, aiming to share one of the most successful leadership experiences in the region, and his insights about the future of youth in energy.

Arranging the Interview

This interview was arranged with a great deal of professionalism that characterizes Sheikh Faisal. To the request, an immediate positive reply was issued by him, who graciously agreed to engage in a conversation to explore his journey towards leadership with us.

We interviewed him while he was about to enter in a Board meeting of Maersk, Qatar. Time was short, but his willingness to contribute to our compilation was huge.

Offshore Qatar

How Did You Started Your Career?

I started as a Petroleum Engineer, in Qatar Petroleum offshore operations. Qatar Petroleum, or QP, is the national oil company of my country, and I was assigned to the operations in the platforms drilling in the Arabic Gulf, one week on and one week off, for two years.

What Did You Learn from This?

It was an excellent experience, which grounded my career with a solid start. I enjoyed every minute of the activities, as I was in contact with advanced technology, and learning one of the core activities of our industry! Additionally, I was trained to follow very rigid safety standards to work on the marine platforms, which imprinted in me a systematic approach to always include high levels of safety standards in operations, which has stayed with me all these years.

You Were in Offshore Operations, but You Are Renowned for Your Insights in Reservoir Engineering and Risk Analysis. How Did You Developed Your Expertise in This Area?

It required time and effort. After my tenure in offshore operations, I joined the Reservoir Team at QP, and worked on building what is called a dynamic model, a simulation model of the reservoirs for two of the most important Qatar's oil fields. I built these models, with the reservoir team with an interactive simulation program, completing this endeavor in two years, 1991 and 1992. These models were used for all the 1990s and early 2000s, as a base for forecasting our production and establishing our operations plans ahead. I consider this was my first major accomplishment in the oil industry.

The Planning Experience

Then, I was shifted to join the contract and planning team of QP. We started by developing the 10-year strategic plan for QP. This was the first time that a modern vision for the development of the oil industry in Qatar was drafted with vision for an accelerated plan of activities, focused to ensure sustainability through time, with growth in production and inclusion of a massive plan for technological applications.

This was a major success, of which I am particularly proud, as it was an eye-opener in regards to the projects, plans, and future of the energy sector for my country, Qatar.

A Sheikh with a PhD

In 1995, I started my Master degree studies in the United Kingdom, which I completed in 1998. Then, I joined the University of Leeds, and in 2003, I completed my doctoral studies, with a PhD in risk management in the oil and gas business.

Were You Only Dedicated to the PhD Studies?

No, I was working at the same time.

During my PhD studies, I joined the International Oil Company operating in Qatar named ARCO at the time. You may know that ARCO was later bought by BP, so we may say I joined BP Qatar. I was with them for four years. This provided me not only with experience in advanced technical workflows, but also provided me with the opportunity to learn more about leadership and dealing with people. I learned what success means in your life. I had the great opportunity to learn all these elements of success working with BP, a major company in oil and gas.

I want to share a memory, from 2000, when I was seconded to work in the UK as operations manager in North Sea. It was amazing how they operated the field. It was a model of excellence in operations for me. A model I wanted to apply in Qatar offshore operations. This struck me, particularly with regards to the efficiency, that I was determined to bring that way of working to Qatar. I wanted to transform the Qatar offshore operations upon my return, to reach that level of efficiency I saw in the North Sea.

I Want to Follow Your Own Realization of Transformational Changes Towards Leadership. So, Could You Tell Me if During that Attachment, Did You Realize You Had Become a Role Model?

Yes, during the second year as Operations Manager in the North Sea, I started to receive feedback from younger engineers who would look up to me as their leader. But they were searching even beyond leadership, for the sake of receiving direction.

They were looking to me as an example, asking for my insight on various matters and seeking my guidance. My point of view started to become guidance for young professionals. I realized I was already a role model. Thus, you become more conscious of what you say to the young engineers, especially the new hires.

What happened next?

In 2004, I joined Anadarko Qatar. It was a middle size company, operating and producing from several oil fields in Qatar. I worked with Anadarko until 2008.

A Fulbrighter!

Sheikh Faisal. I Read that You Are an Alumni of the Fulbright Program. I Must Say I Keep Finding Surprises in Your Trail of Accomplishments. How the Fulbright Came to Be Part of Your Story?

I have always been interested in giving back to the community. One of the Programs that appealed to me was the Fulbright Program, which has an incredible outreach all over the world and is aimed to elevate education through collaboration across cultures.

There is an American Cultural Center in Doha. I received information from this Center about the Fulbright Program, and I applied for the fellowship program, and was selected.

I joined the Fulbright program while working with Anadarko. In particular, I joined the program in USA, as a Fulbright fellow for the University of Colorado at Boulder. I was a lecturer there for the Master program students. For more than year, I taught about Risk Management and beyond the university students. I lectured for the oil industry and the banking sectors.

I was absolutely proud of becoming a Fulbright Fellow. It enriched my vision on the ways of giving back to the community on a global scale. It was particularly rewarding to be able to contribute to the education of the young generation of students in STEM careers, those dedicated to Science, Technology, Engineering and Mathematics.

The Program provided the opportunity to the students to learn not just from the school, but from other leaders. It was an amazing experience indeed.

The Books, the Author, the Famous Expert

It was during those years that I initiated writing about risk management. I published my first book, with my esteemed co-author. I am really proud of that.

Then, other books followed, with co-authors, in a series that has provided me with a healthy sense of pride and accomplishment, as I see those books have been used as texts in universities like Manchester University, and in New York University.

One of the books was translated into Chinese, and is now used in several universities in China. It is a successful book!

What Are Your Books About?

They are about risk management. They are mainly about how you use the risk in oil and gas, and in the financial sector. The structured thinking to face and make decisions which implies risk. How do you do it? The book summarizes a structured approach to the decision-making process. And we also developed a program, and that software is sold attached to the book.

Has the Book Added the Allure of Celebrity to You?

I will tell you an anecdote about my books.

Sometimes, I go to a reception, and some people approached me with "I have an opinion about your book", or "I have a question about your book". This makes you proud.

Then you were promoted to the C-Suite.

The C-Suite

That happened when I joined Maersk Oil Qatar as Deputy Chief Operating Officer, CEO. In this role, I am responsible for all aspects of production and especially for all of the employees. It is quite a responsibility, of which I am very proud.

What is the production of your fields?

We produce about 300 thousand barrels of oil per day. We are the largest operating and oil producing company in Qatar, with 2,000 employees. Our operations are entirely offshore. We have a super-giant oil field, which is a very challenging oil field. It is a multi-reservoir geologic setting, with low porosity and low permeability. Hence, we have wells with multi-completions.

We had built already 15 offshore platforms, with 300 km of pipelines, framed by a 25-year production sharing agreement with QP. They are very challenging operations.

We have accomplished numerous achievements, which are the result of extraordinary teamwork among the Management and the Employees.

In July 2017 the French company Total, will come to operate the field, and we are currently in the hand-over processes. The field operation grew so much! Maersk is a mid-size company and will now focus on its on-going operations in the North Sea, Kazakhstan, Angola, and other countries and regions of the world.

What are your plans for the immediate future?

My tenures with BP Qatar, and Maersk Qatar were an attachment, so I have always been a representative, an employee of QP. At this point, a new role will be in my immediate future. I will take part in defining it, and soon, it will be announced.

Reflections on Success

We Generally Relate to Success in the Presence of People that Have Reached a High Role in an Organization. Or Those Who Have Make Lots of Money. There Are Several Measures of Success. What Is Success for You?

This was the most beautiful and probably the most interesting segment of the conversation with Sheikh Faisal. It was a moment when he shared his insight about leadership, and we discovered that his vision of success is a recognition that the path is not always smooth. We learned that helping others to reach their best helps us to reach our best, as well. His sharing was a moment of authenticity.

For me to be successful is to enjoy what you are doing and to never stop learning. Learn from others' experiences, and learn about their success and failures. I want to emphasize this: success is to never stop learning. To listen, more than talking. I am convinced that education and learning from others makes you successful.

Success also requires one to be in optimal shape in terms of health, which boosts our capacity to deal with stress and huge responsibilities. I consider it crucial to exercise and have a healthy life. Give time to your family. Then, turn problems and challenges into opportunities.

Do not create enemies. Be a model, not a follower.

The Importance Having a Vision and Volunteering

Sheikh, Your Recommendations Are Certainly a Recipe, But One that Is Not Easy! Which Is for You the Most Important Element in Your Recommendations. In Other Words, Which Is that Key Element of Leadership and Success that You Consider Essential. So Essential, that You Would Provide it as a Recommendation to the Young Generations in Oil and Gas. Please, Tell Us

To have a vision!

Have a vision, make a plan of what you want to do in your life, and execute multiple tasks. Fill your schedule with many activities. Relevant activities.

Additionally, I found out that volunteer work makes your professional life and your life more rewarding and also helps in building your leadership style. So, I highly recommend that it's important to dedicate time for volunteering work.

What Has Been Your Own Experience with Volunteering?

I started volunteering when I was in high school. I have always volunteered to do charity work.

Also at the university, there was lots of volunteering work. That volunteering work is important for a leader. The volunteerism gives you an indication of your willingness to insert yourself in leadership. It teaches you how to lead, to coordinate.

During the last 30 years, I have chaired more than twelve non-profit organization like the Qatar Science Club, the Society of Petroleum Engineers' Qatar Section, the Qatar Photographic Society, the Qatar Equestrian Club, Qatar Diving Club, and so many more! And I have learned new strengths in my leadership from each one of these experiences. Most importantly, I felt I was truly helping each of those organizations to envision goals and to achieve them. Most particularly, it provided me with the opportunity to help many individuals to grow to their best. And at the same time, this fostered my own growth.

Which You Consider to Be Your Biggest Challenge?

I think managing people is not easy. This has always been a challenge for me.

You have to understand the minds, the expectations. And that is a challenge, for anyone in leadership. Sometimes, the expectations that people have do not match what you may provide them.

Other challenges I experienced in my career were during the adaptation periods to other countries. My adaptation to the UK, and to USA. I would say that it was a good experience at the end. It was a challenging experience that strengthened me. I could learn that each company has its own style, I tried to grasp the best of each one. Especially in leadership styles.

Qatar 2030 and the Youth

In your opening address in one of the numerous events in which you are invited to speak, you said that "The recruitment, retention and development of Qatari nationals is of utmost importance to Maersk Oil. We are firmly committed to providing the best career development opportunities for nationals, supporting the country's future development under the National Vision 2030." Besides supporting with the best plans in the company, what other message would you like deliver to the young generation of engineers?

The Future in Energy and Oil and Gas

What Do You Envision for the Future of the Energy Sector?

I fully support the vision we developed, which is called "Qatar 2030". It is about a clean environment. It is about a transparent society. I feel that I own the vision in Qatar 2030, and would like to see it is effectively happening in the near future.

We live in an advanced society when you have young leaders in the leadership roles, in leadership positions. See the composition of leadership, and generally, you can detect a progressive organization.

What Is Your Insight for the Future of the Oil and Gas Industry?

Oil and Gas, especially in this region, always goes through cycles. There are cycles of ups and downs in the oil price, and also in the investments and levels of activity. We are now facing cycles of a reduced timeframe, with 2-year cycles. It is very challenging.

I consider the future of oil and gas is based on technology. We have to use technology to our advantage. Technology is our key to propel future. We have more advantages than disadvantages in the oil and gas sector.

A Shared Selfie

- **Your favorite word:** Achievement.
- **A city:** Mecca.
- **An important person of your preference:** My mother.
- **Your favorite food:** Lobster.
- **A favorite music:** My preference is fixed on the music of the 60s and 70s.
- **A favorite landscape:** the Rocky Mountains. I loved hiking in Colorado Springs, in Boulder, at Red Rocks Park, and in so many places! The sight of beautiful sunsets in the Rocky Mountains is imprinted in my eyes.
- **Your favorite color:** Blue.
- **Who supported you the most?** My mother.

Dr. Ramona M. Graves

"My future had to be outside Dannebrog".

A Glimpse

There are people who innovate with every step they take. There are people who easily motivate everyone around them. And there are people who push you to be the best you can be. Ramona Graves is a special kind of person who encompasses all these kinds of people in one being.

In 2012, she was appointed Dean of the College of Earth Resource Sciences and Engineering, the first Dean of the new college and only female Dean on the campus of the Colorado School of Mines. This school with worldwide prestige is one of the oldest institutions in the United States, founded in 1874, and ranked the top engineering institution in the world (College Factual in 2016). Hers is not a minor accomplishment, in this era of empowerment of women, as the Colorado School of Mines did not have a woman in a leading role this important in its more than 143 years of history.

The choice of picking Ramona for one of the highest roles at Mines becomes an obvious one when one reviews her pioneering work in academics. She was the first woman ever to obtain a PhD in Petroleum Engineering at the Colorado School of Mines, and the second woman to get a PhD in Petroleum Engineering in the United States, and one of the first 10 women in the world to accomplish this feat. She was the first to launch and teach integrated subjects dealing with matters of applicable interest in the oil industry (1990), the first woman appointed head of the Petroleum Engineering Department at Mines (2007), and the absolute driving force behind the renewal of the Petroleum Engineering Department of the Colorado School of Mines (2013, Marquez Hall). A true trailblazer in academics, her impact is felt in every remote oil field on Earth, where a petroleum engineer from the Colorado School of Mines works.

She has taught an estimated 12,000 students over her career as professor, in more than 350 courses at the Graduate and Undergraduate levels, and she has supervised 500 thesis students. She led the effort of revamping the petroleum engineering departments in Kazakhstan and other academic institutions around the globe, in what was a developing effort with SPE, to step up the quality of petroleum engineering teaching techniques, with direct interventions, producing immediate and practical improvements.

Ramona is an accessible giant. One who is always cheerful, and does not believe in hierarchical distance. Stepping out from her busy agenda to reply by email or in person is an approach that comes naturally to her. She keeps in touch with her busy network in the oil and academic sector with her unique personal touch and style. She is a tornado of positive energy and passion for good quality education in petroleum engineering, and one feels blessed to have been touched by this strong wind!

- She was born in Dannebrog, Nebraska, USA.
- Graduated in Mathematics and Physics, Kearney State College, May 1973.
- PhD in Petroleum Engineering, Colorado School of Mines, May 1982.
- Petroleum Engineering Department Head, Colorado School of Mines, 2007.
- Dean of the College of Earth Resource Sciences and Engineering, Colorado School of Mines, since 2012.

- 5 Patents.
- More than 95 invited presentations, spanning 5 continents.
- Outstanding CSM Faculty Member-Department of Petroleum Engineering-selected by graduating seniors 8 times since 1995.
- Doctor Honoris Cause Honorary Doctorate from the Mining University of Leoben, Austria (2006).
- Society of Petroleum Engineers Distinguished Member.

A Personal Snapshot

One of my first classes at the Colorado School of Mines, while pursuing my Master degree, was "Integrated Reservoir Characterization", and was given by three professors, one for each discipline involved: Geology, Geophysics and Petroleum Engineering. The purpose was to help the students reach an integrated understanding of the container that holds oil and gas, which in the energy industry is called a "reservoir". In the collective imagination, reservoirs are subsurface pools full of the black gold, whereas the reality is very different, as the oil and gas are contained in the pores of rocks, trapped by impermeable seals, which requires a full characterization, in order to be efficiently produced.

I met Ramona on this learning journey, and I like to say that she is the person responsible for instilling in me the curiosity, passion and love for petroleum engineering.

Arranging the Interview

In 2016, I was at the Annual Technical Conference and Exhibition (ATCE) of the Society of Petroleum where we met to directly capture her stories of leadership. She gave an easy and prompt "yes!" to my request. After a few aborted attempts on my side, partially due to her busy schedule at the event, we finally sat down to initiate a conversation that brought revelations, a few tears, and an emotional sharing of her candid approach to life.

A Tornado from Nebraska

A Farm Girl from Nebraska

Ramona, Tell me about your career.

Tell you about my career?! Tell you about my career?! It has been decades and I have not even tried to summarize my career! This is a good question, and one on which to reflect.

I grew up in a town of 350 people in Nebraska. There were, there are, and probably there will be literally only 350 people in my hometown. The name of this small town is Dannebrog; it is a farming community, one of those towns deep in Nebraska, where you feel lucky to have your own zip code, but feel isolated from the world. I studied my elementary and high school years there, with only about 15 or 20 other people attending classes.

I Have Never Been Good with Blood!

15 people? I cannot imagine what that was like.

Yes. 15 people. Maybe less. It was a very small town indeed. I only had three career options: to be a teacher at the local school, to be a nurse, or to be a wife. We were a farming community, and these were the options available to women. I realized I did not like teaching, have never been good with blood, and did not have many prospects for a husband. So, my future had to be outside Dannebrog.

With this vision, I went to college in a town about 45 miles from home, Kearney State College, enrolling in Mathematics and Physics. It was a town of approximately 20,000 people, and I started to move up to larger and more crowded settings. I graduated with a Bachelors degree from Kearney in May 1973, and then got a job as a high school teacher in Omaha, Nebraska, a city with 350,000 habitants in 1973.

The world expanded. In Omaha, I was teaching math and physics for a high school. You were surprised when I told you I did not like teaching, but this is the work I found, and very soon, I was to discover I could do other things.

Productive Golf

I was playing golf with my date at the time on a beautiful course located in Ponca City, Oklahoma. The place was associated with the refining headquarters of Conoco, and when we first drove into the town, for the first time in my life, I smelled the characteristic odor the refinery gases sometimes bring to the air. I immediately wondered "What is this smell?!! Ugh!" It was my first encounter with the petroleum industry.

You had never smelled the refinery odors before?

No! How could I have, if I always lived and studied in Nebraska. Oklahoma was a whole other story. But let me continue my story. The wish to play golf was very strong, and the place very beautiful, so, we went several times, networking with the players there, who were largely Conoco engineers, since Conoco was the major employer there for the oil sector.

After a few games, and as we became more acquainted, they started encouraging me to pursue an engineering degree, and in particular, a chemical engineering one, as that was the degree pursued by all other women who at that time worked in the oil industry. That was the entrance into the oil industry for women.

This golf conversation motivated me to register as a part-time student at the University of Nebraska, to become a chemical engineer. And I just loved it! But only for one semester.

Only for ...one semester?

Yes. Maria Angela, In the second semester, I had to take this course to study chemistry—heavy, deep chemistry. It was really an awful course. However, it required a special project.

The Special Project, Life Changer

We were required to prepare a special project related to any topic close to chemical engineering. At the time, there was the Santa Barbara oil spill, and the subsidence in Long Beach, that got a lot of attention in the press and in the news.

The communities started to demand their rights for a cleaner environment, and the oil industry was looked upon as not very eco-friendly. It was the very early stages of the environmental movement, and I was in for it!

I went to the library to search for some petroleum literature. You may imagine that I found very interesting books, and my research enabled me to understand more and more about the need of oil for energy, and not only that, but the stages of oil production: exploration, drilling, processing of crude oil, and more.

I was particularly fascinated by the uncertainties and the risks inherent to the oil industry… and I fell in love with Petroleum Engineering. I was certain that I had discovered the love of my life! I decided to change my major and pursue a Petroleum Engineering degree. With a determination that I felt so strongly for the first time in my life, and that was to lead my actions in my career afterwards, I decided that I wanted to become a Petroleum Engineer.

The Choices

I had to decide where to study and found out that the top engineering schools for Petroleum, at the time, were Stanford, Texas A&M and the Colorado School of Mines. So, I applied to these three schools.

Were you accepted at Mines?

I was accepted at Texas A&M and Stanford. And let me tell you: CSM rejected me!

With these choices on hand, and being a Nebraska farm girl, California seemed extremely modern; whereas Texas was closer to what I considered conventional, traditional, and closer to my education and family background. So, I selected Texas A&M. I quit my teaching job and started packing. I will never forget that while I was packing, my father looked at me, and shared what was to be his last advice. "Honey, I hope you find a husband this time", he said.

Seriously?!!

Yes, he was supporting me in the best way he knew, I am convinced of this; in the best way he could.

Life Has Its Way of Evolving

I did not know you attended Texas A&M. Are you an Aggie, Ramona?

Wait, wait, life has its way of evolving. I was getting ready to go to Texas A&M, and received a phone call from Mines. Somebody had withdrawn and I was in! I could not believe my luck. So this is how I went to Mines! It was my preferred option, as it was closest to home and very traditional. Plus, Golden was a town with 10,000 people at the time, versus College Station, at 250,000; and Palo Alto, around 60,000.

Yes, I can imagine you felt right at ease in Mines from the beginning. Tell me, did you face any major obstacles?

The story is long, but I would like to highlight two major challenges. Firstly, a phenomenal roadblock in front of me was the attitude of a "somebody" (faculty), I will not name, that specifically stated that "I will never have the shame of having a woman get a PhD under my watch".

Amazing. I am astonished. Even if this was back in the 70s, as a strong advocate of the advancement of women in their professions myself, I feel outraged.

You feel outraged listening to it, but I was living it! That completely changed my vision and that challenged me. Do I stay or do I leave? I had to make a fundamental decision, and while I was debating what to do, that man was fired from Mines, and a new person was given that position. I was blessed with a new mentor whose first reaction when he found out what happened was to tell me, "I will do everything in my power to enable you to finish". Many things in my career have been serendipity. The impact that one person can have in your life is immense. I experienced this in my PhD studies at Mines, and let me tell you, it was my first real big challenge ever.

"I Want to Do This Forever"

And what was your second most important challenge in your life? You mentioned you had two you wanted to share with me.

Yes! And you will probably appreciate this even more: After graduation, I wanted to work at a major oil company, but I was offered a position to teach at Mines. As I have told you, so far in my life, I was convinced that I did not want to teach! But, at the same time, literally the same day, I found out that I was expecting my second baby, and it was a high risk pregnancy, for which I needed to stay close to home with very short driving distances.

So, you had to teach, without wanting to do it?

Yes, I chose the job to teach at Mines. The extraordinary environment immediately blew me away. I must tell you (as if you did not know) that, at Mines, the students are all very intelligent, committed, and determined to succeed in their studies. Not only that, but their parents are proud, the labs and classrooms were excellent, and I was given flexibility with my schedule.

After only one semester, after finishing my very first semester of teaching, I went to the office of the Petroleum Engineering Department Head, and told him "I want to do this forever". I felt energized. This energy has never abandoned me since I have been at Mines.

A Woman Dean at the Colorado School of Mines

You have a long trail of success, and in several roles. What experience in all these years marked you the most and why?

When I was asked to be Dean, I responded with a resounding "no". I knew I was doing a good job as department head for PE. I was aware of the positive feedback that I had received, from many fronts, and that was gratifying. I did not think that a higher role, as Dean, was good for me, as I wanted to continue to do what I was doing.

One colleague said to me, "You can do this, and you have to make a difference; it would be selfish of you not to make a difference for the School".

For a woman, it was tough to be told by someone that she is being selfish! We women are natural providers, mothers, and caregivers. By our nature, this is what we do best. I reflected on what my dear colleague said, and had to reconsider.

I accepted. Besides, you are never fully ready for the roles that promote you from your current role.

Amazing trail for a decision that was very difficult. What a challenge!

Yes, Maria Angela. But I want to tell you that another great challenge I faced—actually, my greatest challenge so far—was to raise two children while at the same time building a satisfying career. This balance is very difficult and at times I thought it was insurmountable.

So, was that your main challenge?

Yes. If someone asks me what was your major challenge, which is very easy question for me: My children! I have two children: a boy, Jacob; and a girl, Lacey.

Ramona's face lights up with a special smile. You can tell she is proud of her children.

The Sundry Store

Some people had a special influence in our careers. Who was this person for you?

Oh, I would say without hesitation that special person was my mother. She was a very strong woman. I like to say that she was a career woman. She opened a sundry store, which is like a drugstore, with gift merchandise and pharmaceuticals, etc. In the 1950s, in Nebraska, this is definitely not an easy thing to do. She made my clothes, and she offered this tailor service in town. She was on top of all details of the store, and was active all her life. She still is, although she retired (forcefully retired, at the insistence of many!) when she was 82.

What are your personal strengths that you consider propelled your success?
You get me off track here, with this question!
After a while, and a few sips of coffee, she concludes:
I do think my strengths are my sense of humor and my sense of self. I am OK with who I am. I am at peace with what I have accomplished in life. And I am proud to let you know I have approached life with humor. It was not a fast or an easy process at all. In fact, I acquired that through life. It's a slow process. I felt fully actuated when I was 50. I am not embarrassed to let you know that I feel I keep growing in my sense of self. Every day.

Nothing Keeps Me Awake at Night

What are the challenges the energy industry will face in the future?
Nothing keeps me awake at night in relation to the future of energy. Two physical elements in life are a constant, and those are food and energy. We will always need both of those. Now that I am Dean, I have grown immensely in my own awareness about the importance of energy and its components for the global needs.

Without oil and gas, without minerals, the world would not have evolved to the point we are today, and we will need ever more energy and minerals in the future.

What message would you like to send to the young generation?
The one thing I tell all my students, all the time, is to never select a job or career based on money. You make your selection on where you can potentially make your biggest impact, and where you think you will be happy.

If your work does not make you happy, change it!

A Shared Selfie

- **Your favorite word:** Easy! It is "can"! Like in …I can, you can, we can, they can. Definitively, "can" is my favorite word.
- **A city:** Evergreen, Colorado, USA. That is where I live. And it's my favorite city in the whole world because it's my home, where I have my life. I am very happy in Evergreen.
- **An important person of your preference:** Eleanor Roosevelt.
- **A personal happy moment:** There have been so many!

- **Your favorite food:** I like white food: Milk, cheese, bread, white cake, pasta. Yes, white. All white edible things. I loved them!
- **Who supported you the most in your life:** My parents.
- **A death:** My father's sudden death from a heart attack. He has been gone for 25 years. I still miss him very much.

  At this point, Ramona cries. She tells me that for years she did not have a moment of such a strong emotion like this.

  I respect her time, silently waiting. She recovers her composure, and after a while continues, explaining to me in words I will always cherish: "He never got to see my successes".

It is clear to me that the audiences she has addressed in five continents, in big conference halls in the most important world exhibition centers, or those filled with executives in luxurious offices at the top ranks of academia and the oil industry, lacked that pair of eyes she wanted to thank the most.

Post-scriptum

When I finished my interview with Ramona, we were supposed to take a selfie together, but as usual, people were lining up to ask her questions, to require her signature on some important document, and to request her participation in a future event. I left, certain there will be other opportunities.

The very next morning, at breakfast time, when I was going down in one of the 12 hotel elevators, the doors opened at the wrong floor, and incredibly, Ramona was there, with her luggage, and… no place to step in and join the ride in my packed elevator going down.

She saw me, and said "We never took the picture, Maria Angela, step out!! Step out!" I did not step out, in the frenzy of the moment.

I regret it.

Ali Rashid Al-Jarwan

"A leader must have a vision and persist".

A Glimpse

Ali Rashid Al-Jarwan is the Managing Director Exploration & Production and CEO of Dragon Oil, a privately held and wholly owned subsidiary of Emirates National Oil Company, with producing assets in Turkmenistan, Algeria, Egypt, Afghanistan and Tunisia. Previously, he was the Chief Executive Officer (CEO) of Abu Dhabi Marine Operating Company (ADMA-OPCO) from 2006 to 2016, in charge of the two major areas offshore Abu Dhabi, namely, Umm Shaif and Zakum, which are among the world's largest oil and gas fields.

Mr. Al-Jarwan leadership extends way beyond ADMA or even Abu Dhabi, to inspire oil industry workers globally. As an active volunteer for the Society of Petroleum Engineers (SPE), and most importantly, as the chair of ADIPEC 2006, 2008 and 2012, he is one of the top leaders whom others admire. His colleague seek him out for strategic and managerial advice in all kinds of topics related to the oil industry, and specifically for offshore operations and technical events of global outreach.

He is a well-recognized professional. In 2014 he received the highest recognition of the SPE. Honorary Membership is the highest honor that SPE presents to an individual and is limited to 0.1% of the SPE total membership. This elite group represents those individuals who have given outstanding service to SPE or have demonstrated distinguished scientific or engineering achievements in the fields within the technical scope of SPE. He also received the "Distinguished Membership" of the SPE in 1999, and the SPE "Regional Service Award" in 1992. He has chaired several of the most important international conferences in the oil and gas industry, such as GEO in 2002, the 8th HSE International Conference in 2006, and the giant ADIPEC in 2006, 2008 and 2012, a conference that in 2016 reached the status as the most important conference of oil and gas in the world.

- B.Sc. in Petroleum Engineering, University of Tulsa, Oklahoma, 1979.
- MBA, IMD, Lausanne, Switzerland.
- General management degree from Cranfield, School of Management, UK.
- 1979, trainee in drilling operations, ADNOC.
- 1983 Head of Planning Department ADCO.
- 1987 Manager of Petroleum Development ADCO.
- 1997 Assistant General Manager (Technical), ADMA.
- 2001 Assistant General Manager (Operations) ADMA.
- 2003 Deputy General Manager, ZADCO Abu Dhabi.
- 2006–2016 Chief Executive Officer ADMA-OPCO.
- 2017 Managing Director E&P and CEO Dragon Oil.

He has been recognized by top institutions and organizations at a global scale, and has received leadership and innovation awards from BP and ADNOC.

A Personal Snapshot

Volunteering for the SPE enabled an excellent opportunity for Hosnia Hashim to liaise with Ali Al-Jarwan, while advancing several activities for the Middle East, India and North Africa region, when she was Regional Director of the Society. Hosnia always found in Al-Jarwan that special support provided by sincere professionals who share knowledge and experience, without reservation.

Maria met Ali Al-Jarwan when he attended the very first SPE KOGS 2013, the "*Kuwait Oil and Gas Show 2013*", in Kuwait. This is a bi-annual conference which is the flagship event of Kuwait related to the oil sector. Ali surprised all the organizing committee members with his remarks: He mentored and encouraged us to keep pushing for the best feasible quality event. He told us that ADIPEC, the most important event in oil and gas of the world, started just as a small country-focused technical event in the UAE, based in Abu Dhabi. He supported our organizational efforts, and he told us that we should aim for KOGS to grow its size and outreach. His words were truly motivational, and had a special weight, given his importance as a role model in the region.

Ali Al-Jarwan is a very accessible person. It is reflected in all that he does. To our request of learning more about his success through a focused conversation, he immediately arranged for a meeting. It was a pleasure for us not only to remain captivated by his style once more, but also to enjoy once again his humbleness and accessibility.

The Driving Force in ADMA

"Perfection Starts with People"

Ali, tell us about your childhood in Abu Dhabi.
We were ten children. I have eight sisters and one brother. I was blessed with an upbringing full of love, where my father was an extraordinary caring role model, and my mother was a source of infinite support and love.

A Captain to Learn From

My father had four cargo ships, transporting goods from Africa to India, to Dubai, and to Bahrein. From Pakistan to the United Arab Emirates; in fact, there were many routes. It was the old trading commerce, with dates, grain and fruits. But I must tell you this. The principles of trading are the same in all business. And since I was a small kid, he would take me on board, he was the Captain of his biggest ship. I would look up to him as a Captain, so resolute, so resilient. Life was not easy back then, and I saw his determination in achieving his goals. In his business, he needed to adhere to routes, to be responsible, to save resources, and to manage his time well.

Most importantly, I saw him respecting his people, his employees. He treated them very well, and this was key to his success. In retrospect, I think I have applied myself all my life to treat people well, and I do it as the natural thing to be done, following the best examples you can have in life, which are your parents.

How did you choose to study Petroleum Engineering?

In the high school, I was in love with science subjects. I liked geology. One time, I was reading a magazine; it was an Arabic magazine produced in Kuwait, with an article titled something like "This is the way how the oil is explored in Abu Dhabi". It was the trigger of my passion for the oil industry! I decided to study Petroleum Engineering.

My initial plans were to attend the university in Cairo, Egypt, where many of my high school peers were planning to attend, but I was recruited by the Abu Dhabi National Oil Company (ADNOC), which sponsored a scholarship in Petroleum Engineering and a chance to study in the United States. I went to study at the University of Tulsa, in Oklahoma. I had the chance to study my career in the best country in the world for the subject. I was at Tulsa from 1975 to 1979.

My classmates at Tulsa were from Oman, from Bahrain, from Qatar, and from Kuwait. All of them are now in leadership roles in their respective countries. I believe our education was foundational to our careers and achievements. The bond we established during our study years remains strong, and I am very proud about my colleagues' success.

Young and Single

My first work assignment was with ADCO in 1979, in the drilling and production department. I enjoyed my work every day, earning new experiences, proving myself in field operations. I was young and single, so I had no worries about long shifts or continued work for journeys that encompassed weeks in the field.

Then, after two years actively working as a junior petroleum engineer, I was sent to ADNOC for two years, until 1983, to work as a reservoir engineer, which is a more scientific focus. I had the opportunity to be involved in studies and workflows aimed to optimize production, and to forecast the volumes of oil, gas and water to be produced. It was beyond interesting! I developed a passion for the oil industry.

Upon my return to ADCO, in 1983, I incrementally assumed more responsibilities, as Supervisor of Reservoir Engineering, then Head of Development Planning, and then Manager of Petroleum Development, a position I held until 1987.

What was your next step in the roadmap?

I was appointed Manager of Petroleum Development of ADMA-OPCO in 1987. It was a fantastic experience, with plenty of challenges. I met many people, national Emiratis and expatriates as well. It was in these years that I emphasized actions on talent management, how to develop individuals and teams for the overall success of our company.

The Growth

What do you consider to be a top achievement of your years in ADCO?

Those years were key in my profession in many ways, and I would like to think the major two accomplishments were to launch ADIPEC and to technically support an unprecedented increase in our reserves, boosting our recovery factor after a careful study that demonstrated we were too conservative.

We will surely talk about ADIPEC again later in this conversation, but now, please tell me more about the reserves increase. It must have been an achievement with a tremendous impact.

It certainly had a huge impact, and was pivotal for our development plans in the oil industry of Abu Dhabi. And let me tell you an important reflection: this was a major accomplishment, but it was a result of major challenge as well!

I was the head of the Reserves Committee for both ADMA and ADCO. We needed to re-evaluate our reserves, in alignment with our national government strategic objectives, which encompassed to increase our OPEC quota. As you know, OPEC's quota are based on reserves, and we wanted to increase our production.

We had a very conservative look at reserves, that had to be challenged, considering the new technological means of assessing the resources, and producing the oil. We worked closely with Shell, ExxonMobil, BP, and other international oil companies, in collaborative efforts that were fruitful. We re-positioned not only Abu Dhabi's ADNOC in concert with OPEC, but also enhanced the skills of our specialized workforce related to subsurface studies and reserves certification processes.

Focus on People

Always the people, I notice that you attach all your achievements to the people, to your employees.

At the time, I led all functions related to the Field Development, like geophysics, geology, reservoir engineering, petroleum engineering, and all subsurface sciences. And I recognized once more the value of the people in all we do in the oil industry, and if I may extrapolate, in any activity in life. It's especially true for reaching targets that require collective, integrated efforts.

Let me detail an important achievement. I was in charge of different reservoirs, and responsible for many pioneer activities for the oil industry of my country, such as the first 3D seismic we ran in Abu Dhabi. I was also very proud to be responsible for the first horizontal well we drilled.

When did ADCO drill that first horizontal well?

It was in 1986, towards the end of my tenure. It was a big success, which made us more receptive and even keener to implement new technological solutions that could boost our production. Almost immediately afterwards, we engaged in multilateral drilling, fish-bone drilling, and other new technologies in drilling wells.

During these years I developed a technical insight on subsurface matters and managerial skills, especially with respect to people.

Escalating to the Top

Would you conclude you had a rapid growth?

No. I had a gradual and incremental growth, handling incremental responsibilities, which exposed me more and more to people. Those who worked internally in my department, and the people outside ADCO, our stakeholders, contractors, liaisons, and partners. I became convinced at heart that people are at the center of all our achievements.

What trails of leadership did you develop in this stage of your career?

I would say two: communication and tenacity. What you call resilience. I had to communicate internally, motivating our young generations, and the not-so young as well.

(Here, we laugh. We comment that all people in the teams need the motivation, even the "push" from the leadership to move ahead, upwards or onwards, to achieve specific objectives).

And then, also to communicate and motivate the external parties, our partners and liaisons.

Leadership for me, is a learning journey, one in which you lead others in their learning journey, and you increase your own learnings. It is incremental. Every step of the way, a new learning. I enjoyed listening to all, as every piece of information, and every perspective counts. I realized that a multi-faceted view at any problem brings more solutions than a single, hierarchical, or authoritative approach. And this is true in both technical or managerial realms.

I developed a team approach, in order to achieve things through the work of my teams.

The Executive Path

Then, I was engaged in more-executive roles. From 1997 to 2006, I held three executive positions. The first one was Assistant General Manager of ADMA for all Technical functions. I was leading activities for field development, commercial, engineering and projects, HSE and quality assurance for ADMA, essentially all technical functions of the company. It was a role that is now called Deputy CEO—Technical Functions.

In 2001, I became Assistant General Manager of ADMA for Operations, responsible for the operations of our oil fields Umm Shaif, Zakum and Das Island. I was responsible for the production and export as well. I held this role until 2003. I was taking care of the entire operations and our entire offshore. The third executive role was in 2003, when I went to our sister company, ZADCO, as Deputy General Manager for the entire company for three years, ending this appointment in 2006.

What were the learnings from this first executive period?

The importance of people in all you do. As an executive, this finding was evident once more. People were even more important.

The first of June of 2006, I became CEO of ADMA.

The CEO Responsibility

It was a time to handle huge projects. My task was to develop our production capabilities to ensure a production of 1 million oil barrels per day. It was a long journey, with big investments of multi-billion US dollars. We had to achieve a sustainable production of 600,000 barrels of oil per day, and then add 300,000 more from new fields.

It took ten years all together. It was a long cycle. These are mega projects with multimillion dollars investments. We had to develop fields from concept, from a project on paper to actual production of oil from offshore wells. For example, for each field development plan, the surface facilities could cost more than four billion US dollars.

How many mega projects you had at the time?

I was handling three mega-projects that I steered from the design stage to the engineering, procurement, and construction phases.

And this mega-growth while having to sustain and even increase oil production! What were the main challenges?

The integration of all elements.

I will highlight one of the elements that will exemplify the growth in activity that I led: we increased the rig count in Abu Dhabi from 12 to 26 rigs.

We wanted to do all this within the best standards of safety. I am very proud that when I went to ADMA, the average of safety indexes was 18 million work-hours/year and when I left ADMA, it was 60 million work-hours/year of incident-free operations. Almost with the same workforce headcount, so we improved the HSE mentality and approach. It was an efficient cultural change towards safety. One of my objectives was to reduce activities that were risky. We reduced the severity of incidents.

I took charge of the codes of practice, for example, the procedures of ADMA. One of the different committees that I chaired at ADMA was the one related to ADNOC Marine Standards. Since we applied the procedures, we have had no fatalities in 17 years. When you care about your people, it shows in these kinds of indexes.

Our achievements were distinguished with many awards.

How did you change the corporate culture? That is one of the most difficult things to achieve.

It was not only me. We considered ourselves as one team. One example of this is that we launched the accountability team: people were not delegating (they could not!). Everybody shoulders the accountability. This enhanced the individual accountability on safety matters.

Every year, we advanced one topic, engaging almost everybody. And I mean everybody in the workforce. The first years, for example, we devised the vision and mission for the company. We organized numerous workshops, making sure this was a collective corporate approach, and not a personal, CEO-related issue. Finally, I was managing the entire group ADNOC: Zadco, Adma.

Once the vision and mission were set, we focused on one strategic topic each year as a theme.

The Topic of the Year

One year we explored our journey on excellence. Another year, we targeted the collaborative environment. And so on. Focus enabled people to keep improving, focusing on certain elements.

I remember my favorite themes: Excellence, transparency, respect, and collaboration. There were other themes. With this approach we developed a defined corporate culture for ADMA.

You have focused intensively on building a collective culture and approach to work addressing the whole workforce, changing the corporate culture step by step. What about the individual knowledge of the employees?

That is a very interesting question. Without the individual knowledge, we do not have collective progress.

Their Subject, Their Community

I developed a framework for launching communities of practice. Basically, we were inserting those in the Key Performance Indicators (KPIs).

For what topics?

For everything. For Operational Excellence, for example, we developed a community, inviting those individuals with clear passion for developing and sharing the best practices in drilling. It was their subject, and thus their communities.

Within 3 years, we achieve multicultural communities across the companies, on a variety of disciplines. The communities were across the groups, to give the professionals a free zone to think and adopt the practices. I have done this for five communities: Drilling operations (4 companies), Maintenance

(7 companies), Quality Assurance and Quality Management (9 companies), Integrity Management (9 companies) and Safety Environment (9 companies).

These five communities reported big savings and provided much optimization to ADMA.

"I Was Convinced ADIPEC Would Be Big"

Tell me about the jewel of event that is the Abu Dhabi International Petroleum Exhibition and Conference (ADIPEC). In 2016, ADIPEC welcomed 97,000 attendees, making it the top event in the oil and gas industry in the world. How did it all get started?

I was the founder of ADIPEC. It was in 1984. I was leading local section of the Society of Petroleum Engineers, SPE Abu Dhabi. I initiated ADIPEC as a small conference, and I put loads of energy into continually organizing it every two years. I was convinced this would be big in the future.

How many years you were the Chairman of ADIPEC?

For 10 years. Specifically from 1982 until 1992. The first event was in 1984, but we started organizing it in 1982.

Then, I turned over the responsibility to young leaders to make it grow even more and to take the event to new heights. I was the Chairman of ADIPEC 2006, 2008, and 2012, when it became a super-conference.

This was really a major thing. I hold ADIPEC close to my heart.

Many people participated in this evolution, creating the current success, including the many authors, the contributors, the volunteers, the SPE, and infinite numbers of people.

Every year I was recording the progress. I am still surprised with every new edition's milestones.

Recognitions and Awards

How did you started in the SPE?

I knew about the SPE since my studies in Tulsa. I started as an active member in the Middle East Oil Show (MEOS), back in 1980, as an organizing committee member. Then, I became 1985–86, SPE Regional Director for the Middle East, hence an SPE Board Member.

My volunteer work in the professional Societies has brought me many recognitions and awards, for which I am most grateful.

Who were your role models?

I think I have had very few role models. But first of all, Prophet Mohammed, secondly my father, and third, several important leaders whom I met in my career. One is David Woodward, an executive from BP from whom I learned many things. But I have learned from so many people.

Leadership Reflections

Along your career, when did you feel you were a leader?

It was a process. It was gradual and, I would say, imperceptible. I had a passion for the oil industry, which compelled me to work hard. Then, may be as a result of dedicated work, or as a natural ability, I was able to lead incrementally-difficult processes. Leadership for me was a gradual process, not something that came with a tag or role.

I left ADNOC, ADMA with a good reputation, and I am glad about it, because I worked with my heart. I hope I was a good role model. I did this for my country, for ADMA. And for myself. If you want to be distinguished CEO, you have to be on top of all dynamics.

What do you consider to be your best leadership skill or characteristic?

I would say honesty. It is important for a leader to be honest, and to be transparent.

Another thing I would like to share is that I engaged and keep engaging in continuous learning. You can always do better. Nothing is enough. Learning is fuel for the mind, for your success.

And the perfection in leadership, it all starts with people. Good leadership comes from how you treat people. Create challenge for the people. Be a people role model, be accountable, be with them in good times and in difficult times.

The Patience Needed to Succeed

I am sure you have inspired many leaders in oil and gas, as well as the young professionals who have just joined. What messages do you have for the young generations?

I will direct my comments to our young new hires in ADNOC. I would advise them to be patient.

That it is my recommendation to stick to the professional ladder. Nowadays, the company has system for managerial and technical ladder. For leadership development, there is a process for talent management, with much more clarity than in my years about how to pursue their careers. We created the system for them, for the young professionals. They need to care for themselves, and now there are options. And they will advance, irrespective of difficulties. They are different than my generation.

And what are the challenges you envision for the energy industry of the future?

Firstly, the technology deployment cycle. It should be faster. To do more for less.

Secondly, to encourage innovation in all processes. To encourage people to bring cost-effective solutions. More creative solutions. Especially now, that we have fallen into a low price of oil cycle.

Are you concerned fossils fuels will be displaced?

No, this is the dynamic of the world. We are always creating something new, for a sustainable future. To ensure that people will be able to increase population, they require quality of life, and access to goods and services has to be cost-effective. We will see renewables coming in with strength, but we are still cost effective, and we will continue to be an important part of the business.

A Shared Selfie

- **Your favorite word:** Excellence.
- **A city:** Abu Dhabi. In the Islamic world, Mecca. For international cities, I like Milan and London.
- **An important historical figure:** I like all kinds of people. I like to learn from different people.
- **Your favorite food:** I like meat and fish, but I now tend to eat more fruits and salads.
- **Your favorite color:** Navy blue.
- **Who supported you the most in your life:** My mother. She gave me unlimited moral support and appreciation. And I was inspired by my father.
- **You happiest moment:** to be frank with you, at my age, I have seen the happy and the sad moments of life. I appreciate the good moments, and I always keep the happy moments in my mind. Not a particular one, but all.

Post-scriptum

Ali Al-Jarwan is one of the top role models in the Middle East region. In one of the initial meetings of the Executive Committee of the 2017 Kuwait Oil and Gas Show (KOGS), when a high-profile person as a moderator was needed for one of the Executive Panels, the name of Ali Al-Jarwan was proposed immediately.

Someone said "Mr. Al-Jarwan retired recently". Several committee members immediately replied "retired? No way. Al-Jarwan will never be fully retired. He is part of our oil industry! He will always be active!"

That statement, agreed by all, ended the discussion. Ali Rashid Al-Jarwan was invited to participate in the program of KOGS 2017.

Olivier Soupa

"I wanted to be in the inner circles of the organizations, close to the core business".

A Glimpse

When you need support in the executive workflows related to the optimization of organizational structures, you call a consulting firm. There are many consulting firms, especially for the oil and gas sector. But when you need a good consultant you call him or her by name. Olivier Soupa is one of those advisors who executives in the energy sector call by name.

Olivier has been a key force behind the reshaping, enhancement and uplifting of a few major national oil companies in the Middle East and Africa. His insights and leadership were and continue to be instrumental in developing many of the major milestones of key projects in Kuwait, Saudi Arabia,

UAE, Angola, Nigeria, North Africa and others countries in Europe and the Caspian Region.

Olivier Soupa is a name that typifies discretion in the executive support. He provides consulting services of the highest level, to reshape, in many cases, the organizations of the most important companies in the region, to provide results and roadmaps to be followed towards success. He is part of the leadership team of Accenture Strategy Energy for the Middle East, based in Paris, France.

- High School: Saint Jean de Bethune, Versailles.
- University of Law of Paris (Nanterre).
- Sciences Po Paris.
- Started as a Consultant at Andersen Consulting in 2000.
- Joined Schlumberger Business Consulting (SBC) in 2005, Consultant to Senior Manager.
- Founded and led the SBC Energy Institute in 2009 as Managing Director.
- 2011 Principal of Schlumberger Business Consulting.
- In 2015, SBC was acquired by Accenture, and he became Managing Director for the region Middle East and North Africa.

A Personal Snapshot

We know Olivier from the time he was working in Schlumberger Business Consulting (SBC), a branch of the main giant international oilfield service company, Schlumberger. SBC was acquired by Accenture in 2015. For years, Olivier had provided the support needed to revamp Kuwait Oil Company and KUFPEC's organizational structure and workflows, to reach excellent, world-class levels to rival the best in the sector.

Closer contact and professional respect grew gradually, as the work and projects that were developed by Olivier evolved and produced many fruitful and useful results, supporting transformational strategies, to bring success and ensure the continuity of the changes or improvements proposed.

The most recent collaboration in which we engaged together was the preparation of the international presentations of Hosnia Hashim at the conference PetroTech, and for the Annual Technical Conference and Exhibition (ATCE) of the Society of Petroleum Engineers, in Dubai, during October 2016. It was an amenable experience, with excellent teamwork, in which Olivier's contributions were key. As usual!

The purpose of our interview was to learn how Olivier developed his career, in the back office segment of the business, covered by Consultants. We want to know how he managed, immersed in the leadership discussions, outside of the mainstream, and to understand how he managed to survive, in the rarified air of the executive corporate ranks.

Arranging the Interview

This was one of the most structured interviews of all of the interviews that we arranged. With a very formal meeting request, confirmed and unchanged for any "urgent matters." Olivier arrived at the right place, at the right time, without not even one minute of delay for any unexpected reason. His proverbial punctuality was exactly what one would expect from an expert in consultancy services.

Political Sciences

Olivier, What Did You Study and Where? Tell Me About Your Formal Education

Very seriously, Olivier quickly summarizes his formal education history. I observe he is obviously still not very comfortable with being the focus of this conversation. Soon, he starts to realize his story could be a role model for the myriad of consultants of all levels who seldom get to learn how others succeeded. I am one of those, and my interest in his answers motivate an openness about his own journey. And after very few preambles we start.

I started studying Political Science, to become a public high-ranking officer in France, a very prestigious and quite stable career path. This type of career path is also very slow and full of interweaved difficulties that may accelerate or hamper your career. My university is called Paris Institute of Political Studies, in Paris, France.

Sciences Po, as it is called in France, is an elite institution, ranked 4th in Politics and International Studies by QS 2016 World University Rankings. The school was created in 1872, and the staff include twenty-eight heads of state or government, including the last four French presidents (François Hollande, Jacques Chirac, Nicolas Sarkozy—although he didn't graduate—and François

Mitterrand), thirteen past or present French prime ministers, twelve past or present foreign heads of state or government, and a former United Nations Secretary-General. The alumni include CEOs of France's forty largest companies.

The External Circle

I was destined to join the governmental sector. And I was focused on that idea and objective. But at the same time, I was studying Law, as I wanted to work in a business Law firm. And I did! At the end this was my selected path, to work as an attorney internship in a law firm, taking care of clients needing legal consultancy services or legal support of all kinds.

When I started in a very prestigious Law Business Firm, I realized almost immediately my work was in fact way too far removed from the reality of the action, of the field, of the ground work. I felt too detached.

What Do You Mean Too Detached? Lawyers Are in the Center of Many Kinds of Actions, Aren't They?

I felt detached. Really! Detached. What I mean is that, as a lawyer, you do not see the impact of what you do. You certainly help your clients in a big way, and all companies need lawyers, legal services, but I felt detached.

I see the company as nested concentric circles, of all functions, where lawyers are in the external circle. Operations is the core, the heart of the company, the essence. The company grows and needs support services, and all kinds of organizational structures that fundamentally back and sustain the operations. And at the very far back, in the outer circle, you will find the legal services. So, lawyers, in my vision, are far from people who may directly impact the business.

I had to do something to change my future. I did not want to be in the external circle.

What Did You Do to Change This Realization?

This is the point that changed my career. When I realized what I wanted to do, I struggled for a while. You may imagine it was not an easy decision. I took a step forward and actively changed my career.

Tell Me About How You Did It, Not Many People Have that Kind of Courage. Especially at the Beginning of Their Careers

I joined the consulting sector to be closer to the center of the company and our clients: I joined Anderson Consulting, which was the cornerstone that became Accenture. At the time, it was a nice strategy consulting firm, very pure in the philosophy of work, not an IT-focused company.

The Key

In consulting firms, you do not know with whom or in what you will work. Any given project rules the focus and rhythm. Actually, you do not need to be a world expert, but you have to understand the overall dynamics of a sector and have a certain depth in operations. For the rest, you just have to work, study, and analyze the challenges at stake with facts and analytics. But work faster. Work faster on the items, to understand things and …solve them. This is the key.

I wanted to become someone who can talk about a specific topic better than others. I was working in several sectors, so I decided to specialize in Oil and Gas (O&G). I liked the sector, I gained more and more experience, I was (and am!) convinced the energy sector moves the world, as no other sectors brings together such an intensity of technology, geopolitics and investment. Therefore, after a few years, when Andersen Consulting shifted clearly towards IT, I moved to Schlumberger, in a consulting firm focused *on* the sector I was interested in.

This is how I joined Schlumberger Business Consulting.

From the Champs Elysees to an awful Crappy Hotel in Angola

The shift of work was also a shift in my life, moving from Champs-Elysees in Paris to an oil and gas guest-house in Luanda, Angola.

A Tough Move Indeed!

Yes, you cannot imagine. My office overlooked one of the most magnificent views of Paris. Every single day, I could see through the window the glamourous images you only find in the Champs Elysees. An elegant style and focus at work suddenly changed for what was to become the norm in the coming years of my profession.

Were You Alone at the Time?

No, I was already married. It was a difficult decision not only for me, but for my family. An agreed decision. I was always rotating, so my family was not with me. When I started, I spent 18 months, a full year and a half, on a rotating scheme of three weeks of work in Angola, and one week back home.

What Role Did You Have?

I started as a senior consultant, and quickly became Manager. It was a rough business, dealing in those years with the Angola O&G sector, with many players involved, but I gained momentum. Most importantly, I gained the trust of our clients, delivering what they needed, and flagging their gaps to be mended, so it was a fruitful relationship that provided my company with opportunities to keep growing its reputation as a reliable consulting firm.

"I Started to Behave like Owner"

When Did You Feel the Wakeup Call to Leadership?

The wakeup call. What a nice term in relation to leadership!

Olivier here sits back at the chair, and I can see he is mentally shuffling options, considering several memories and peak moments. His gaze shifts, and finally he declares the right moment when he realized he was immersed into leadership, becoming a leader himself.

Yes, it was a realization. A wakeup call. To me, that happened in 2009, when I was asked to create a special group for SBC. It was called the SBC Energy Institute, and from this moment onwards, I started to 'behave like owner'.

The Owner?

Yes. You start believing you do not work for an ever-changing, incognito client. Or for your employees. There is a moment when you realize you are not working for someone. You start to work as if the whole thing is your company, your baby. You start working for specific objectives, for a goal, a specific target.

It was this feeling that made me feel compelled to steer, to motivate and to guide all the teams, and the stakeholders, into the same direction, to optimize every single bit of the business in the Energy Institute. Caring for it. Day and night.

Participating in the creation of a new business unit within a larger company—in particular, within Schlumberger—is probably the most important and impactful experience in my career. There was a unique spirit of entrepreneurship in the team, coupled with a strong sense of belonging to a main player of the oil and gas industry. It was at the beginning of the oil boom in 2005, and it was recognized as one the strongest successes in management consulting in the 2005–2015 decade.

Never on the Front Seat

And as a Leader in Consultancy Services, What Does Satisfy You?

Consultants are never on the front seat. They remain as low-profile advisors and stay in the backstage of the clients they serve. Therefore, our achievements are those of our clients and the impact of our projects are visible in the field.

You must know the saying among us consultants: that the clients say *Success is thanks to me, and failure is the fault of Consultant.*

How many times I have been in that situation myself. I can now see Olivier has shared that kind of perspective, and multiplied by the number of employees in his teams all over the world or the region he is handling in any specific moment in his career.

In my particular case, I feel recognized when I know I have gained the trust of the executives of those companies with which I deal. Especially when a client liaises with me, appreciative of the services or consultancy rendered by one of our team members. In that moment, I know I have developed in our team members that sense of belonging. Perhaps there is also that sense of ownership that drives leadership in their inner selves, guiding them to a sincere desire to deliver the top quality service. This is noticeable and highly recognized by the clients.

I remember having high levels of satisfaction attending a ceremony celebrating the acquisition of a company, which was the result of a complex project on which my team and I had worked days and nights during weeks and weekends. That ceremony was the visible celebration of a series of agreements, studies, discussions, workshops, meetings, and analysis and roadmaps that finally became a reality for the management, for the operations and for the employees. It was a peak moment, when as a consultant, with an owner mentality, I was absolutely gratified by that special joy of a work well done. The special joy that big achievements bring to you.

I kept applying the perspective of being the owner when I left the Energy Institute. The time heading the SBC Energy Institute was a good family time. It was good not to travel so much, because we were having our children, and I could fully enjoy their early years including the first steps and the first school reports.

After that, I wanted to go back to pure consulting projects; hence back to a commuting scheme. I guess I am a seeker of change, for the very nature of my consulting job, always advocating for change, always fostering uplifting maneuvers.

I am the Father of that!

I went back to the core consulting business in 2011. I came back to the consulting business leaving behind the great experience of the SBC Energy Institute, to become a Principal, reporting to a Vice-President. This new role enabled me to attain and kick-off a series of projects and business development activities of which I am absolutely proud.

What Has Been Your Major Achievement so Far?

With a modest smile, Olivier looks at me, mentally reviewing the many successes in his portfolio and CV, and after a few seconds, he picks two.

There is a study in SBC, which has become the SBC HR Benchmark, which is presented every year at a global scale, generally attached to the yearly HR Forum event, in Paris, that had great resonance among all major oil and gas companies.

I Know About the HR Benchmark! The Second Slide of the PetroTech Presentation We Just Built Together This Last Month Came from that Benchmark Report, Correct?

Yes, it came from that report. I have led and delivered this initiative during seven years, from 2005 to 2012.

It is a well-known annual statistical and trends report that deals with the major topics useful for Human Resources. The report had an impressively relevant impact. I still receive all kinds of positive feedback about it.

The Global Warming

The other achievement I would very much like to mention is the study on Global Warming, a project that was developed as part of the key initiatives of the SBC Energy Institute with a few talented young consultants I supervised.

In 2007, with Al Gore's Nobel Prize, the Topic Started to Receive Serious Attention. Your Study Echoed that Trend?

Yes. This initiative within the Energy Institute was certainly arriving at the right time. Now, with the 21st annual Conference of the Parties (COP21), the strategies related to the conservationism of our environment are again at the center of governments' and companies' roadmaps. We will all have to migrate to this new framework in the Oil and Gas sector. It is going to be difficult, but it is a ready-set-go kind of approach, as there is a clear need to swiftly implement agreed strategies to protect the environment.

Olivier, These Are Fantastic Achievements, Not Only for You but for Your Teams. Indeed. I Want to Know Now, …What Do You Brag About?

Believe it or not, I am proud of the job of Consultant. If I were to brag about something, I would brag about that. I know it is a bit in the shadows, but I feel that it is thanks to the consultancy work that companies and organizations in many sectors thrive and escalate to improved levels of efficiency and leadership. And I am proud of that.

Is There Any Particular Client's Case that Made You Extremely Proud?

A few projects we did for several national oil companies in Sub Saharan Africa fall in this category.

I was happy to see how these companies have evolved over time, how they have moved from being non-operating partners to being large operating companies in charge of complex oil and gas fields. In great part, this progression towards national empowerment of operations was the result of our work as consultants, through a series of complex and delicate processes of asset transfer. I am definitively very proud of how much was achieved by our dedicated commitment.

The Excellent Speakers, the Story-Tellers

Tell Me About Your Role Model or Role Models

I have known many extraordinary leaders and role models in my consultancy job. As I liaise with executives, in an ever-increasingly frequency, more and more I enjoy the opportunity to admire CEOs in their moments of leading decisions. In different regions of the world. I have always been attentive to people sharing a vision, for whom the sky seems to be the only limit, who have the strength to change the course of things, despite adverse circumstances.

I have been also influenced by some clients who show a remarkable capacity to make quick and important decisions, showing great clarity of thought and strong oral communication skills. For me, as a consultant, the story-telling capacity of a leader is foundational. Good speakers who express well-structured ideas are rare.

But There Must Have Been Someone Who You Admire the Most or Remember the Most

Michael Liebreich is one of the best speakers for delivering messages to top executives. He is founder of Bloomberg New Energy Finance, and an Advisor for the UN's Secretary General for the "Sustainable Energy for All" initiative. I consider him to be a role model because, when you listen to a lecture from him, you do not even have to take notes. He is a story teller, and, for me, he is so clear. The structure of his speeches is so crisp and clear. I wish we all could reach that communication capacity. Amazing!

Let Me Explore Your Inner Self Again by Asking What Are Those Personal Strengths that You Consider Propelled Your Success?

I will reply to you in a nutshell: work capacity, synthesis and rigor in analysis, and sense of service.

What Are the Challenges the Energy Industry Will Face in the Future?

Here, the executive consultant in Olivier takes over, and he expands with detailed explanations of what he envisions for the future.

I will focus my response on the future challenges of the oil and gas industry rather than the entire energy sector, because I think each energy type will face diverse challenges. I would like to emphasize on four challenges that may be game changers:

1. the intensity of technology,
2. the potential peak of demand,
3. the efficiency of oil and gas companies, and
4. the capacity of the sector to attract the best talents.

Cutting-edge technologies are already part of the energy landscape, and this will grow in the future as oil and gas becomes increasingly difficult to find and extract. Protection of the environment will play an increasingly important role.

Like New Technology for Fracking?

Correct! As we see today in the United States, fracking for shale gas tends to use less polluting products injected at lower price. A paradox appears here, with the challenge to maintain large capital investments to finance costly innovative technologies, while oil price is low and external funding is shifting towards renewables.

Cyclical Turbulences

The need to consider demand for oil is a new challenge, which will become more critical in the next few years as we observe a potential peak of oil demand. The O&G exploration and production sector has always considered demand as being able to absorb the entire hydrocarbon production. We are used to cyclical turbulences like today, where demand is lower than expected because of the global economic outlook, but we are not used to a world where oil and gas consumption structurally decreases.

The challenge of the peak of oil demand will force oil producers to consider the needs of consumers, their locations, and how to market oil. For O&G companies, it will mean an increasing importance must be given to commercial capabilities.

The third challenge I would like to highlight is the need for a stronger efficiency. We remain below other heavy industries' standards in terms of cost management and overall efficiency of process execution. The O&G industry has gone through a decade of booming growth, which resulted in inflated capital and operating costs. We are capable of developing huge industrial projects, but we need to learn to do more with less.

Paving the (Bad) Way

And the Fourth Challenge?

You keep me on track here! Well, the fourth challenge I perceive is perhaps more perturbing: the loss of attractiveness of the oil and gas industry to recruit talent.

Tell Me More. I Am Worried About Our Huge Industrial Sector Losing the Capacity to Attract Fresh Graduates

Me too. The facts tell us that with the oil crisis, many companies have cut recruitment. Leaving the campuses is extremely harmful for the industry. We should continue to recruit and prepare for the future. We are currently paving the way for the same situation we faced during years 2005–2010, when the competition for talent was very fierce, driving to inflation of salaries and many questionable recruitments in terms of quality. Large North American universities will survive, but many other smaller academic institutions may not.

Not Disappearing Anytime Soon

Olivier, You Are Detailing Your Vision of the Future. Now, Please Share What Message Would You like to Send to the Young Generations?

First, that the oil and gas industry is not going to disappear anytime soon. The development of renewables is a significant trend, but in the future energy mix, fossil fuels will remain one of largest energy sources. Therefore, young people starting their careers now will have plenty of long-term opportunities to feed a rich and long career.

The current oil crisis is a difficult moment, but recruitment and career opportunities will resume. I am sure!

Second, the younger generation cannot abandon the technical disciplines. We need engineers who are passionate about their jobs, and O&G offers unique opportunities to exciting and rich careers. Third, energy leaders need to make it clear that oil and gas companies have to reinvent their managerial models.

Reinvent Managerial Models! I Would like to See that Happening Soon

I think we all agree that the time of the old traditional large petroleum company, with a *one culture fits all* model, is ending. Large oil and gas firms will become energy companies developing their activities beyond fossil fuels. The size and duration of large E&P projects and the specialization of technical disciplines are also adding complexity to large organizations which are forced to promote diversity in career exposure in order to retain talent.

How to Achieve that New Model of Management? How Do You Envision This Transition?

IOCs and NOCs will have to collaborate more and more. The boundaries between operators, oilfield service companies and Engineering, Procurement and Construction Companies (EPCs) are also changing rapidly.

The need to develop value locally will force localization of industrial work. All these aspects are great challenges for the young generation, who will have to rethink the way we operate one of the most critical industries in the world.

In other words, the energy industry has never been so full of opportunities to reshape the world and make it better.

A Shared Selfie

- **Your favorite word:** Idea.
- **A city:** New York. And Paris.
- **An important person of your preference:** A teacher, from the years back in Law School.
- **A personal happy moment:** Every single time I return from a business trip and land in Paris, I like the calm of Paris streets at 6:00 am and arrive at home for breakfast with my wife and children.
- **Your favorite food:** Saint Nectaire, a French cheese, with a glass of Burgundy.
- **Your favorite color:** Navy blue. All kinds of blue, but navy blue is my favorite color.
- **Who supported you the most?** My wife. She has been there for me all along.

Post-scriptum

We followed up repeatedly with Olivier for the photos for his section. As a consultant, it was fun to find out he did not have a formal recent picture taken. The back seat is not a place where many photos are taken. We manage to obtain one, and it shows Olivier with the serious attitude he usually impersonates. He is a leading Consultant in the consulting business. It shows. The other photos show his softer side as a family man and adventuring deep in the desert sands of the Middle East, as part of the enjoyment time of his most recent business leadership experience.

Intisaar Al-Kindy

"My turning point was to realize I was courageous and strong".

A Glimpse

A young Omani woman taking over the huge task of launching a new Shell office in Jordan is certainly an unusual image, even these days. But that is exactly what Intisaar Al-Kindy accomplished, and this accomplishment took place more than 10 years ago, back in 2007.

Intisaar al Kindy is one of the pioneering women leaders in the industry. She is a pace setter in Petroleum Development of Oman in that she is the first Omani national to hold the role of Exploration Director in more than forty years of history at the company. She possesses many such records, and she is a role model in Oman and the GCC for the new generations of oil industry workers, both men and women as well.

A natural instinctive leader of huge stature, Intisaar approaches life with the determination of a truly courageous innovator who has no fears about challenges ahead. She has demonstrated that she is an absolute master of work-life balance. The many facets of work come along those of life, and if they do not match as a synchronous opportunity, she will make them match successfully, through the choices she makes.

- Master of Science in Geology, from the Imperial College, 1990.
- Geologist from University of Tulsa, 1988.

A Personal Snapshot

The first time I heard the name of Intisaar Al-Kindy, I attempted to pronounce it properly, using several extra consonants that my mind included where there were none. It turned out the name had a very simple pronunciation, that the linguists would express like iy-NTiy-SAA-R. Intisaar is an Arabic name, which means Victory, Winning, and Triumph. What a fantastic and premonitory choice of a name for a person that personifies just that: Triumph!

We had invited Intisaar to be the keynote speaker for the Kuwait Oil Company's first women-matters conference event, launched in 2002, during a time when very few organizations were openly addressing women matters in the oil industry. She delivered an impeccable keynote speech in front of a large audience, explaining how PDO encouraged women engineers to work in operations.

With a spontaneity that captivated all, and a genuine, candid, and direct style, Intisaar was the first oil-sector leader (men or women) whom we heard directly emphasizing the need for quotas. She stated that without quotas for imposing female participation at work and at leadership ranks, it would not be feasible to obtain quick wins, to enhance the gender diversity in the oil industry. Needless to say, Intisaar gained our admiration …and our hearts!

Arranging the Interview

Intisaar accepted immediately the proposal from Hosnia to be interviewed in Muscat, capital of the Sultanate of Oman, where Hosnia and I planned to meet her. But after rescheduling several times, we opted for a phone interview. Intisaar graciously accepted and we initiated what will be a journey of discovery of success and leadership of this amazing woman. She declares she is just a hard worker, in a humble statement that matches her life-long stance at life and work..

When the call commenced, she initiated our discussion with a hurried "I am so busy, the last three weeks, I have had no weekends. This is the first time I am taking whole contractual accountability for a big, huge, immense project that we are about to launch in Oman. But happy to be here, talking to you." Like the true leaders we have come to know, Intisaar dedicated a full, concentrated attention to our objectives and ultimate goal of the interview. She encouraged us to pursue our idea of capturing the human side of the leadership stories of our contacts, and with that genuine motivational words from this magnificent role model, we started.

The Junior Geologist that Grew as a Giant

I Am Just a Geologist!

Intisaar, how did you become such a famous leader in Exploration and Drilling Engineering?

I am none of that! Even today, I hesitate to call myself a renowned expert in drilling or a know-it-all in exploration. I like to repeat to all that I am a geologist. I am just a geologist by my academic background. I have worked all my career in the oil industry, and I initiated my first step on my career ladder by working as a junior Geologist in PDO.

At the time, I already had two children, because I married and had them while I was studying for my bachelor degree in geology. From my joining day in PDO, the main company of my country, I started on my career path in the company. My work was conducted in several different roles as a geologist at the office and at the field, and was mostly related to PDO's production workflows related to oil and gas exploration and production, our main resource.

Life Happens

But then, life happens to you, and I got divorced. I wanted to keep working, in my usual work, but PDO had other plans for me. Instead of continuing a regular career like other colleagues posted in Muscat, PDO wanted to send me as a seconded personnel in Shell in London, to be engaged in seismic interpretation. I received this news with surprise, as at the time I had six years of experience, but also five years as a mom of three children, and was a single mom.

Oh! And what did your family and your friends advise you to do?

From the beginning, I wanted to accept the challenge, but you cannot imagine the myriad of criticisms and warnings I received. Yes, I faced all kinds of severe and acidic criticism and especially warnings for my choice of accepting the challenge, to engage in this assignment with my children, alone. Some of these warnings were based on caring for me, for my future, I realize that now, as a mother myself, but others were just criticism for unusual choices. Courageous choices.

They would ask me things like *Will you have enough money? Who will take care of the three kids? How will you take care of their schooling and your work at the same time, as a single mom?* But we survived, and we actually thrived as a family. And me? As a professional, I learned a lot from the technical point of view, but most importantly, I gained a new perspective at life and work that I will cherish forever. I realized self-confidence is based on my own learnings by doing, by trying, by attempting even—especially—the most difficult things.

In a Deep, Blue Sea

After finishing this assignment, Intisaar came back to take her first appointed leadership role in PDO. She had already been a leader in what she did, and set an example for many in PDO with her assignment in London, but I can hear she is willing to tell me the details of this phase of her career, as this was the first time she was formally appointed a leadership role in PDO organization.

When did you finish your assignment in London?

I came back to Oman in 2001, appointed Team Leader of a small team, with which I grew as a leader until 2007, when Shell selected me to start my biggest challenge to date, which was to go to Jordan to set up Shell Oil Shale company in Jordan.

An Omani national to set up a new Shell office in a new country?!

Yes.

That was a challenge, because not only I did not have experience in setting up a company; I am just a geologist who drills wells. But it was a complete new territory. I felt I was just dumped in a deep blue sea, to either swim or drown.

I had to start everything from literally scratch. Even today, I consider it was that job which made me into who I am today. PDO and Shell recognized how I made something out of nothing. I like to think (and several executives have confirmed that to me!) that I was named the Director of Exploration because Shell decided that if I survived there, in the middle of nowhere, in Jordan, launching a new Shell office, I could easily be the Director of Exploration of Oil and Gas here, in Oman.

So, you did not drown in that deep, blue sea?

Intisaar laughs for a while, and her laugh is indeed contagious. We both laugh, and it is a wonderful moment. I imagine a young geologist with this huge assignment, and I wonder, how did she manage?

No. For a woman who was very junior, with nothing to offer but passion for what she does, believing in herself, and wiling to reach out for help, it seemed an unsurmountable challenge. I could not imagine how PDO was so confident in selecting me. Deep inside me, I knew the size of the challenge was huge.

The Terror that Shapes Who I Am

(I paused a while, getting the courage to ask the following question to this impressive woman) … Intisaar, were you scared of this new assignment in Jordan?

Are you kidding me?! I was terrified!

This answer completely disarmed me. I was expecting the usual … "I became courageous, the common "I felt very confident on my capacity" speech we generally hear from leaders of her stature, but as I came to learn, Intisaar is your down-to-earth kind of leader, who will describe for you the facts as they are. No more, no less. Then, she completely opened a window into her inner-self. I will remember the details and tone of this portion of our conversation for a long time.

Until today, I think that no one knows that I am very good at portraying a self-confident image, but it is this terror that shapes who I am, because when I am doing a challenging job, that frightened feeling gives me the energy to succeed. I always ask myself: What is the worst that can happen? Most of the time, the answer is that nothing happens. Nothing catastrophic, nothing irreparable, nothing definitive.

I fully believe that 95% of my terrors do not materialize. So, yes, you can be scared about something, and even so, you will succeed. Or, like in my case, thanks to terror you will succeed. Even failure will teach you and help you in your professional growth.

This characteristic is very unique to women. I would say that many times we women internalize our fear, and we can find all the reasons for something not to work. We say: *I will look like a fool if something goes wrong, this cannot work out,* and *something will go terribly wrong.* We have a thousand reasons like these. We have this disk inserted in our head, which makes us not have courage. So I challenge you to think *What if?* And you will see: Nothing. Nothing happens. As long as you are good in wearing the mask of confidence.

Showing No Fear

Intisaar shares her soul in the following remarks, and opens with me the core of her leadership.

For me leading is showing no fear, while delivering excellent performance. For me, it is OK to be scared, but keep it inside you, because your shareholders want to see and hear a confident person.

It is fine to experience physical symptoms, physical consequences out of fear, because it happens also to me. But as soon as I stand in front of my audience, may that be a small group of fresh-graduates, or company shareholders, or the full-house Board of Directors, there is a confident person in front of them.

The Turning Point

When did you realize you were a leader?

When I got divorced. Full stop. When you have nothing to lose, you have to lead. We go through milestones in our life. These milestones will direct your life to unfold either in support or against you. When this major milestone in my life happened to me, I thought I could not support myself and my three children. I definitively realized I could lead.

At the time, everybody was warning me. *Are you crazy?* they asked me repeatedly: *"Go to London with 3 small children alone?" "Recently divorced?" "Is not better to stay in this difficult time with your family?"*

So I assessed my scenarios, and I realized if I could not handle it, I would have collapsed anyways, whether I was in London or in Muscat.

Not many women have the courage to take risks. We are raised and educated to be always accompanied, and to be led. Especially in my culture.

When my colleagues were asking *Are you alone? Are you a single mom? Are you here working and studying?* I was given the gift of taking calculated risk. And to succeed at it. I suddenly realized I was very courageous. It happened in London. While I was alone, with my three wonderful children, working, and doing the household chores at the same time.

Many women do not learn to appreciate themselves. Instead, I say, actually… I scream to them: "NO! You are a shining star". You have to tell that to yourself. You have to absolutely convince yourself that you are. It all starts from that point. Convincing yourself. My turning point was to realize I was courageous and strong. At that point in my life, I was ready to take on anything.

What Diameter Does the Well Have at the Bottom?

Your turning point came outside your country of origin, Oman. That is remarkable, to say the least. What else did you gain from this first assignment?

The assignment to Shell in London provided me with the gift of what I call permanent memories. Memories I will always cherish. During my assignment in London, I was given extraordinary opportunities. For example, I met King Abdulla and Queen Rania, of Jordan. And I met Queen Elizabeth of England.

Really? The royalty of two countries! How did that happen? Tell me all about it.

My encounter with the Queen of England came as Shell UK was celebrating 100 years of Shell in the UK. Shell wanted to portray a company which is very diverse and inclusive. I was selected to give a presentation to the Queen about Shell Exploration in the UK, and the value we were adding. I represented the two kinds of diversity Shell was striving for: gender- and nationality-diversity. And I had the added bonus that I served in a technical role and could explain the exploratory plans.

I started my presentation in front of this big audience, in front of Queen Elizabeth II. At one stage, she asked me what the well size (the diameter) was.

The Queen wanted to know the diameter of the exploratory wells?

Yes. She asked just exactly that. And I of course felt at ease; I was in my element! I replied the diameter of the casing varied from top to bottom, and it is smaller at the bottom. She nodded and said thank you. She showed a genuine interest in what I was presenting to her. And I can testify that the Queen of England now knows the diameter of the wells drilled for exploratory in the North Sea.

Shell trusted me to present for her. And that was just great.

Were you representing PDO or Shell in that opportunity?

I was seconded to Shell UK during that time and represented Shell.

A Gift

All my assignments have strengthen me beyond my expectations and imaginations. The learnings I built from these assignments is that we, women, most of the time do not realize our gifts. The moment you discover your gift you become a different person. And you can play that card in many ways.

What is your gift, Intisaar?

My gift is being able to communicate and engage in a way I make people believe I am a very strong person who delivers.

But you are!

Yes, but I do it with a team. All my achievements have been the result of teamwork.

The One-Person Office

Tell me more about your job in Jordan.

Shell wanted to set up and run an oil shale business in Jordan. That is it. That was the goal. And I was given the role to launch an office in Jordan for Shell, establish all workflows, create legal standing for the company, hire the personnel, and deliver the whole package.

Amazing! You must have felt proud.

I was proud. This came about as a result of the vision of a senior person in Shell UK who knew me and had the courage to believe in me, and in my capacity. I landed in Jordan with a determination to make it happen, and I learned on the way. I learned how to launch a new business, from dealing with real estate agents to arranging for an office, to recruiting a multi-disciplinary and integrated study teams that would assess the venture. I was proud, yes, but more importantly, I was glad that the objectives were reached.

If You Want to Talk About Added Business Value, You Will Find My Name Next to It

Of all your achievements and challenges, which one was important for you?

I love your question. Maybe because I love the most recent achievement. The recent one is icing on the cake. It's about embedding the mindset of profitable growth in explorers' ways of working and exploring for new concepts. What drives explorers is going after big reserves volumes. The bigger the discovery, the better. The challenge was to explore for volumes that added profitable value. I have pushed this mindset and transformed the exploration group in PDO. Many oil companies in the world consider Exploration as a "research unit", not as business unit. In Oman, in PDO, we have changed that. PDO Exploration is adding value to an extent that is monetizing our discoveries in a record time. Let's see what happens in the last five years in PDO. If you talk about added business value, you will find my name next to it.

Exploration value added comes from fast hook-ups and quick monetization. The time from discovery day to development is shortening dramatically.

Two Kinds of Role Models

Some people had a special influence in our careers. Who was this person for you and why?

I would consider there are two distinctive aspects when you talk about role models. There are those who inspire you spiritually, and with spirituality you gain courage. Others, give you tools to enhance your skills.

The role models who push you in your spiritual sphere, and in the end, they push you to learn more. And to learn by yourself. My grandmother was a very special grandmother. A very special kind of person. Because her perspective of life. The question I always ask: *what is the worst that can happen?* came from her, a person who did not write or read. The person who ultimately built my character was my dear grandmother.

Intisaar then tells me about her second role model…

Another role model from which I received what I will call a "pull" and visible advice, was Ceri Powell, the current Executive Vice President Upstream International Exploration of Shell. She is the one who recommended me for the job in Jordan. She recognized that I was not fully ready, maybe just 60% capable, but she took the risk, and she sponsored my name to be selected. She pushed me to my best, enabled this opportunity for me, and I had to raise myself up to fit into the role. She really has promoted national talent in the Middle East, envisioning, selecting, sponsoring, championing. She is my role model because I have started to do what she did with me. So, if I see a person who is 60% ready to do the job, I will take the risk of the remaining gap of 40%, and appoint this person to the new role.

There Is so Much to Be Harvested

In my case, Ceri could have easily picked a man to do the job. There were more capable men around and more experienced than me, but she took the risk with me, an Omani woman. I am repeating what she did with me. I have been able to keep promoting the national Omani talent to reach their best, making me and all of us proud.

No person is never fully ready for the role. Especially for women, this is a fact, because we tend to think less of ourselves. You will never know unless you put them in a situation of "sink or swim". If they are 60% ready, you can rest assured, I am going to provide the needed support. I tell them I will be only a phone call away, to help them grow and develop the missing 40% of skills or expertise.

There is so much to be harvested! When you take somebody who is 60% ready, not only you are giving this person an opportunity to prove themselves, but you are building trust and respect.

What are the challenges the energy industry will face in the future?

Let's first understand the young generation. It is a generation that wants to be rewarded for waking up in the morning. They do not want to build the credibility that comes with hard work. The good part is that they believe in themselves, and they have enormous energy and drive. The bad part is that you have to convince them to do the jobs that will help them build skills. How to balance between harvesting the creativity the millennials have, and motivating them to sustain performance, that's the challenge.

Energy Is Not Only Oil and Gas

The future is a hard reality; no doubts we will see a transformation towards renewables and towards another source of energy which are not oil and gas. The next generation of leaders will have to be creative, skilled in both depth and breadth to be effective. What we have today is a generation that is challenging; they want to be on a driving seat before getting a license to drive. That mentality of *I am ready* needs to go away. Credibility comes from understanding the reality of our countries, and understanding that energy is not only oil and gas. They need to build skills, networking, reading, learning, to be first in technology. But I get worried when I see them coming to work in the morning, and completing some small task, and asking: "I did this project, I want to be promoted!!" I do not think they realize promotion to leadership role is earned. It is not a right. I remain worried when I see these attitudes.

What message would you like to send to the young generation?

Never stop learning, focus on building your credibility. I have never asked for a promotion. I earned it. My delivery and performance spoke for me.

A Shared Selfie

- **Your favorite word:** Authenticity.
- **A city (your favorite city in the World):** London.
- **An important person of your preference:** My Grandmother.
- **An important historical or religious figure of your preference:** No doubts! Prophet Mohammad.
- **A personal happy moment: important historical or religious figure of your preference:** The birth of my daughter. I love children, and to tell you the truth, I was hoping all my children would be girls. My two first children were boys, and when my daughter came, immediately at the delivery I sat and asked: where is my girl? Where is my girl? Show me it's a girl! Show me! I did not let the nurses take her away.
- **Your favorite food:** Anything with lots of sugar in it! And I am a chocolate person.
- **Who supported you the most in your life:** I supported myself. My inner support is my best support. And I am blessed with 6 sisters. They are always there. With their solid and strong moral support. That kind of support you need to accomplish something.

- **Your favorite color:** Green! No hesitation. All my things are green.
- **A death:** My grandmother passing away. She passed away ten years ago, and I feel the pain of her loss like it happened yesterday.

  She taught me that important question in my life. The question of "what if", and made me realize that nothing catastrophic will happen, that the World keeps spinning, and the Sun comes out every day. Every single day!

Post-scriptum

In response to our request of sending us several photos that would show her at work, Intisaar sent a variety from which we selected this one. Her inquisitive nature is shown here, and it is truly a characteristic of this energetic lady, who obviously does not leave any detail to the random flow of events, but captures the opportunities with passion for success. Because, "what is the worst that can happen?!"

Dr. Behrooz Fattahi

"I always wanted to stay in the applied side of things".

A Glimpse

Volunteers in professional societies are the very heart of the organization, shaping the activities, the governance, and its future. Some volunteers are best at propelling change, envisioning new futures, establishing liaisons and bridges where there were none, and advancing the membership to new levels.

Behrooz Fattahi has been one of the most energetic forces in the Society of Petroleum Engineers.

In a career that extended over 37 years of dedication to the oil industry, and continues to flourish with many successes, Behrooz has developed a recognized expertise in heavy oil. He was the Heavy Oil Development Coordinator of Aera Energy LLC, an affiliate of Royal Dutch Shell and ExxonMobil, involved in developing the Company's heavy oil assets in Southern California. Behrooz is a compendium of knowledge in reservoir engineering of heavy oil. He has taught technical courses in topics related to reservoir engineering and enhanced oil recovery for Aera and for the industry worldwide.

But the shining profile of Behrooz comes from the many accolades he has accumulated serving several of the top professional societies and, in particular, the Society of Petroleum Engineers (SPE). Behrooz has a list of achievements so long that it will serve as a role model for volunteers for many years to come and will be very difficult to emulate. He is currently the Vice-president of the SPE Foundation. He previously served as the Executive Editor of the SPE Reservoir Evaluation and Engineering Journal, as the Director of the Western North America Region on the board of the Society of Petroleum Engineers International (SPE), as President of SPE Americas Inc., and as SPE Vice President-Finance. He was the 2010 President of SPE, and, as such, chaired the SPE Board of Directors.

His volunteering capacity has also led him to serve as a member of the United States National Petroleum Council and as the 2014 President of the American Institute of Mining, Metallurgical and Petroleum Engineers (AIME). Indeed, the volunteerism of Dr. Behrooz Fattahi creates a persistent wave of improvement that impacts all people around him, and that motivates young professionals to increase their participation in volunteer roles at SPE. Prior to joining the oil industry, he conducted research for the National Aeronautics and Space Administration (NASA), and the National Science Foundation (NSF).

- Bachelor of Science in Aerospace Engineering, Iowa State University, 1970.
- Master of Engineering in Aerospace Engineering, Iowa State University, 1972.
- Doctor of Philosophy in Aerospace Engineering and in Mechanical Engineering, Iowa State University, 1976.
- Advancing within Shell and Aera Energy, eventually retiring as the Learning Advisor for the company.
- 2010 SPE President, Society of Petroleum Engineers, a society with 168,000 members worldwide.

- 2014 AIME President, American Institute of Mining, Metallurgical and Petroleum Engineers.
- Recipient of AIME/SPE International DeGolyer Distinguished Service Medal, Society of Petroleum Engineers International, the top distinction to recognize distinguished service to SPE, in the profession of engineering and geology, and to the petroleum industry (2016).
- SPE International Distinguished Member (2016); Continuing Education Service Award (2007); Technical Editorial "A Peer Apart" Achievement Award (2007); Editorial Review Committee Award of Appreciation (2008).
- More than 60 Publications in journals, and prestigious international conferences in the oil sector.

A Personal Snapshot

The first time we heard the name of Dr. Fattahi, it was already accompanied by a prestigious reputation, as in 2008 he was already a strong nominee for SPE Presidency. He was elected for the period 2010–2011, and as in 2009 Ms. Hosnia Hashim was selected as the SPE Regional Director for the Middle East, North Africa, and India region. Their liaison evolved to become a fruitful cooperation. Hosnia expanded the activities and actions of SPE in the Middle East and North Africa, in tight cooperation with the leadership of SPE, at the time presided by Behrooz.

Behrooz continued to grow in influence within SPE, leading several committees and key initiatives, some in cooperation with Hosnia. During this period the first Task Force was charged to examine the issues related to women in the oil industry.

Our wish to interview Behrooz was grounded in the tremendous impact his volunteerism has had on Hosnia's contributions as an executive in the SPE ranks, and a role model that he personifies for so many in the oil industry.

Arranging the Interview

Behrooz lives in the beautiful state of California, USA, so it seemed natural to schedule an interview with him in person during the 2016 SPE ATCE Annual Technical Conference and Exhibition (ATCE) in Dubai, close to our work and home, Kuwait. We knew he would attend this ATCE, to receive in person his DeGolyer Medal and it seemed like a natural opportunity to sit together and interview him in person.

It was not an easy task, as Behrooz's agenda was already set and programmed full, with few empty slots weeks prior to the conference. But when luck meets preparation and good initiatives, things have a way of flourishing. I was able to interview Behrooz on a fantastic afternoon of October, in the World Trade Center of Dubai, in the United Arab Emirates, over coffee. It provided an opportunity to conduct an open conversation about work and life.

From Iran to Iowa

Tell us about your career

I was born in Teheran, Iran, nearly 70 years ago. Yes, Maria, that was completely another era in Iran, I am sure you are aware of this. My family decided to send me to study in the U.S., and I have been in the USA for more than 50 years now.

Behrooz's words immediately draw my attention to the fact that he does not explain with detail his reference to Mohammad Reza Pahlavi, the last de facto Shah of Iran, and a status that enables all to refer to him simply as "the Shah". As a Latin-American, and of a younger generation than Behrooz, I become immediately aware this extraordinary individual in front of me grounds his life in a unique, different context than mine. I become even more interested and engaged in profiling him well.

Apollo 11, 20 July 1969

Behrooz, where and what did you study?

I studied and received Bachelor, Master in Aerospace Engineering, and PhD degrees in Aerospace Engineering and Mechanical Engineering, all from Iowa State University.

As an Iranian, I was moving counter-current, as I did not imagine myself to be part of the oil industry. Instead I intensely wanted to be part of the aerospace industry. It was something that I was personally seeking. I was very much attracted to the space program. It was fascinating for me.

How did you become so interested in Aerospace?

I must admit I was motivated to become involved in space matters, in aerospace, much earlier on, from my elementary school years. I was inspired by the enterprises of Yuri Gagarin and Alan Shepard. I was really interested, and I read all materials about them and their endeavors in the space travels that would fall into my hands. That area of my academic education was my personal most important goal: I wanted to achieve something in aerospace.

But… You now work in the Oil industry. How did that happen?

When I was finishing my Bachelor degree in Aerospace Engineering, the Apollo program reached its peak, with the landing on the Moon, and almost immediately, in very few years, the NASA programs started to ramp down, and people started to get laid off. So, I expanded to Mechanical engineering, as a necessity. I had to find a job. I was already married and a job was needed!

The Call from Shell

In the final years of my schooling, along which I was involved in teaching, I started looking for job opportunities appealing to me, and matching my academic profile.

I had specialized in computational fluid dynamics. The aerospace industry was far ahead of other industries in fluid dynamics simulations at that time. The oil industry became interested in simulation modeling, actively looking for people with that skill.

They wanted to find professionals with some background in this field. So I got a call from Shell. They explained to me they were interested in getting experts into the oil industry to work in compositional fluids dynamics, to better understand the reservoirs. They wanted me to go to Iran, to participate in creating a computational model for the naturally fractured reservoirs in Iran. So I did that! The offer was very good compared to a professor's salary, and I joined the team to write the codes for establishing compositional models for the huge oil reservoirs of the carbonate rocks in Iran.

Great! At the time, you were probably one of the very few pioneers building compositional models. So, why did you change?

After a few years, I realized my path would make me a pure mathematician, and I wanted to stay on the applied side of things. I became a reservoir engineer in order to use the models. The company I was working for was a joint venture of Shell, and 12 others in Iran. After the Iranian revolution, we came back to the States, and I started working for Shell in New Orleans, Louisiana.

A Car Ride of a Lifetime

It just happened accidentally. I was riding in the car, with the Shell Division Manager, and the conversation flowed very agreeably into different topics. I mentioned my son was about to go to college, and in his target study area, the schools were not so good in Louisiana. I mentioned I really wanted to go to California, for my kids to attend the top schools there.

Six months later, the manager was transferred to California, and he asked for me by name to transfer to Bakersfield, as he remembered our brief talk in the car from months earlier. It was supposed to be a short assignment for 3 years, I ended up working there for 25 happy years. At first, I worked for Shell's Cal Resources; a company to operate California fields, and founded by Shell. Three years later, it merged California division of Mobil Oil, becoming a new company, called Aera Energy. I had a senior technical role, working on heavy oil projects.

My Children, My Priority

I stayed for 25 years in Bakersfield, and my son and daughter both went to UCLA. My priority became our children's education. The transfer to California was important to me, due to my family's wishes, and my desire for our children to attend the best schools they could attend. I put my family ahead of my career.

What did they study?

After undergraduate study at UCLA, my son attended, and eventually graduated from Columbia University in New York. After undergraduate work at UCLA, my daughter attended law school at the University of San Diego, and graduated from there with a law degree. I am absolutely proud of my children. They were and continue to be a priority in my life. I managed my work so that I could be near them, supporting and guiding them.

At this point, Behrooz looks at me with a proud smile on his face, with shiny eyes, full of happiness. He is indeed a proud father, I think, and I realize this is the right overture to talk about success.

A Successful Package

You have a long trail of success, and in several roles. What experiences in all these years marked you the most and why?

Success is a relative term. Some people may not go to school, but then accumulate tons of dollars, which is not my brand of success, but are considered successful by some. Success for me is a whole package. Obviously money is a part of it; but for me, family comes first. If I have a home that is happy, I can translate that into productivity at work. My model was that I wanted to engage in a job to which I go happily, and when I went back home I would be even happier. I feel I am a very lucky person.

Every day I am adding to that happiness, that enhances my productivity, with 50 years of a happy marriage, and two successful kids. This is my package; the one I built for myself is indeed a successful package.

The Replacement

One of the things all admire in you is your dedication and success to advance positive progress for SPE, the Society of Petroleum Engineers. How did you start your activities with SPE?

I always volunteered in my community, wherever I was. But it was not in a structured approach. I volunteered for the usual Parent-Teacher Association (PTA) and community activities in our neighborhoods. One day in Bakersfield, someone called me and ask for my support in the local SPE Section, as he was being transferred to Houston. He asked me "Would you like to do that? Please say "yes", pal. I have to leave next week, and there are a bunch of things going on."

I had no idea what kind of volunteer time, skills, networking or ability my saying "yes" would entail. I was introduced to SPE in early 1990s strictly by chance.

I am a true believer that opportunity comes to everyone, but it is up to you to miss it or build on it.

I remember vividly my first opportunity in SPE. The section was offering two short courses every year. The year after I took over the section, we delivered 30 courses! My section made 200,000 USD in the next few years, a section that before would just break even. We started a program of scholarships, giving the proceeds back to the students, and to the community. SPE and volunteerism is about giving back. I was thrilled!

I Have to Put My Own Mark

You evolved rapidly in SPE.

Yes, success to me means to create opportunities and build on them. I started as Section Continuing Education Director, then to Secretary, Treasurer, Chair of Section, member, and Chair of the SPE International Continuing Education Committee, and Regional Director for SPE Western Region. Then, I became the SPE Vice President-Finance, President Elect, and eventually President of SPE in 2010.

Of all your achievements and challenges in your volunteerism for the SPE, which one was important for you?

In all roles, from the lowest to the highest, the most important thing is to make a difference. If an opportunity for me will mean to maintain the status quo, I do not take it. I have to put my own mark! I have done it in my company and in everything I do.

Three Role Models Lost in Time

Some people had a special influence in our careers. Who were these people for you and why?

I had 3 people who influenced me, with a long lasting impact, besides my parents:

1. A professor at Iowa State University, Dr. Stevens. I took a computer programming course as a sophomore. He was teaching it, and I got an F on the first quiz. Obviously, I did not understand the computer programming! So I went to tell him I would drop the course. When he asked why, I said I was not in the habit of getting an F. He invited me to his office that afternoon, and for three long hours, he taught me what he had taught in class for that month. Making sure I understood, at the end he gave me a simple problem to do overnight, and that if I did it, I should remain in the course, because that meant I was understanding and progressing. He spent his private time with me to make sure I got back on the right track. I came out of the course with a B, after averaging all straight A's with that terrible F. That professor made sure I succeeded. He is a role model for me. I want to mentor people in this way. Wow! Every time I remember this, I am so thankful. Dr. Stevens was one of a kind.
2. A Professor in graduate school, my academic advisor for my PhD in Aerospace. Dr. Hsu. He told me the door to his office was always open, and made me feel comfortable. I happily went to his office, not missing

any of his office hours. He was always available, and he always had interesting things to share for me to learn more and more. He was very encouraging.

3. Dr. Helmuth Niko who was my first supervisor when I entered the oil industry, at Shell. The interesting thing was that I came from aerospace engineering, without any idea about the fundamental concepts of porosity or permeability. I was thinking, "How can Shell justify paying me, if I do not know the fundamental concepts in the petroleum science?" He gave me the book "Applied Petroleum Reservoir Engineering" by Craft and Hawkins, which is a classic reference! He told me that if I had any questions, I should go back to him and ask questions without hesitation. "How much time should I take for this?" I asked. Dr. Niko mentioned that he did not have much use for me if I did not know the fundamentals to become useful for the oil industry. He helped me with the exercises in the book. In a few weeks, I was ready, and he gave me my first project. So, at critical times in my career, these people appeared on my career path, accepted me, although at times, I knew that I slowed down their work. But they expended extra effort to develop me. That is why I easily progressed to a mentoring role later on in my life. I encourage anyone to go to their local schools, mentor, and share their knowledge with young professionals in their companies.

37 Mentees

I have mentored 37 people; 25 of them are from the oil and gas industry, and 12 from different industries. They all were referred to me by word of mouth. There is an interesting case of one mentee that came to me recommended by his mother. Yes, his mother contacted me, asking for an opportunity for her son. She was from California, and we hired this young fellow as a summer intern in my company. I hope I will be as inspiring for them as my mentors were for me.

I like to emphasize that for me, people come first and then, the job. If I recruit you, the things that happen to you as a person are as important to me as your productivity, because these two are very much related. As long as you stay reasonable, I will develop an empathy for you as an employee.

"I Am Organized"

What are your personal strengths which you consider propelled your success? Behrooz looks at me as the question is very usual, and without hesitation he answers with a declaration.

I am organized. I do not want delays to be a result of procrastination. I request respect for deadlines and workflows. When I was leading the SPE, an organization with 70–80 million US dollars, with about 170,000 members, 400 employees, it was like being the president of a corporation. You have to be organized to handle this kind of organization, because there are too many details and activities.

What are the challenges the energy industry will face in the future?

I will reply about the Petroleum Industry. The Energy industry is a much larger scope, so I will only refer to the Petroleum industry.

The petroleum industry has always been challenged in terms of the technology to apply, in terms of social responsibility, and in terms of providing energy for the world. Now, these are not always aligned. Sometimes, you have to sacrifice a bit in one of these elements, to gain on another. The easy oil is disappearing, forcing us to focus on the difficult oil. If we do not care about the environment, people, countries, and governments will not let us shift to those more difficult places! We have to provide energy for our planet, and we are responsible for our survival in the long run. But we cannot completely forget about the environment. The approach of "I can do anything I want, because people need me" is absolutely wrong. So, you have to come up with a plan to satisfy all the concerns together to go forward. On the top of the challenges I already mentioned is another task, and that is to find talented people to tackle these difficult challenges.

Then, there are other energy resources that are out there with which the petroleum industry has to compete.

What message would you like to send to the young generation?

When I started to work, it was all about technical abilities. Today, we live and work in multicultural communities. When you work in Kuwait for example, you need to understand other cultures around you, and they need to understand you! As a result, soft competencies have become quite important, as important as our technical knowledge. If you are not able to work within a team to which you are assigned, or convey your ideas properly, you will hurt your career.

Storytellers

You have to balance both: Best technical and soft competencies. I always bring this example. Imagine I want to learn tango. Someone can teach me this in a few minutes. The dance steps are not that difficult, but this is the technical part of it. The soft part is the art of implementing it by making an impression on the audience. That is the communication part of the tango!! If I just go there and do the few steps, audience will not be impressed. What the professional dancers do is communicating their techniques to you to move you. They communicate emotion. So, you see? Anyone can learn the steps to tango. Few can use tango to deliver the emotional message.

I can work on field development projects. But if I do not communicate the goals and the vision properly, I will not be successful. This is one of the most important and desired abilities in today's business: communicate the goals beyond the coldness of a black and white bulleted list, transforming those into a colorful vision that moves people into ownership, and into action.

You were the initiator of the Soft Skills committee at SPE. Tell me how you managed to activate this within SPE.

It took me a full year to convince the SPE Board. Even as President, it took me a bit of time to convince them. I kept trying until they approved it. Now, it is a reality, not only here in the USA, but all over the world, there are workshops on soft skills organized and endorsed by SPE that benefit people in the oil industry. I am very proud of this particular accomplishment.

A Shared Selfie

- **Your favorite word:** People.
- **A city (your favorite city in the World):** Paris.
- **An important person of your preference:** Albert Einstein. I find him to be an extraordinary person who remained a human being, down to earth, making a big impact in our daily lives, staying humble.
- **A personal happy moment:** When I got married. Specifically, when my wife said "yes!" after I proposed.
  His face was clearly denoting his mind was not anymore there, at the Dubai Trade World Center, but far away, in time and space. So, I waited in silence, and after a while, he spontaneously continued.

I proposed to her in Iran. We were neighbors, you know? We met when we were kids. I was 14, she was 12. We grew up. And we did not know that we loved one another. We were just neighbors. I left to go to the U.S. to study. It was then that we both realized we were missing each other. I thought about the past, and I realized I missed her so much. I decided to propose. I talked to my parents, and I found my parents were open to that idea. And they supported me, even in the middle of the studies.

I proposed, and, to my surprise, she accepted. Her father was against her getting married to any student, since he believed that a student's future was uncertain, but again to my surprise both her parents came in support. That was the happiest moment of my life.

- **Your favorite food:** I have many favorite foods! But I would say it is pizza. I like the typical American Pizza. Vegetarian. The thick one.
- **Who supported you the most in your life:** My wife!
- **Favorite Song:** "Yesterday" and "Imagine", by The Beatles. And "What a Wonderful World", by Louis Armstrong. One by the Bee Gees and Celine Dion called "Immortality".
- **Where is Home:** The entire world is my home. I do not lie when I tell you I see the map borders as truly artificial lines. Whether it is ethnicity, language, border, color of skin. All of these are artificial dividers that make me feel that I am not at home. I feel that I could go to any place on this planet, and adjust and adapt.

Post-scriptum

When I finished my interview with Behrooz, we agreed to meet up again at the SPE Awards Dinner that night. The photo here shows the moment when Dr. Behrooz Fattahi received his AIME/SPE DeGolyer Medal, a prestigious recognition given to a small percentage of SPE world membership, and the AIME 2016 Presidential Citation from the American Institute of Mining, Metallurgical, and Petroleum Engineers. The picture includes Nathan Meehan, 2016 SPE President and Nikhil Trivedi, AIME President.

His smile and humbleness was present that night, as any other day. He would become a role model of even higher stature with these awards.

Fareed Abdulla

"I was determined to be a Petroleum Engineer!"

A Glimpse

There is one event in oil and gas that grew from a country-scale technical conference to become the largest conference and technical exhibition in the world. That is the Abu Dhabi International Petroleum Exhibition and Conference (ADIPEC), and Fareed Abdulla is a key engine propelling ADIPEC towards success in its latest editions. It is very difficult to imagine the modern ADIPEC without thinking of Fareed, as in 2016 he was the technical program chair. In this capacity, he steered the profile and content of the event, liaising with an amazing network that he handles with knowledge and grace, all over the world, especially in those countries for which oil and gas is a primary export product. The executives from the Middle East, Europe, North America, Africa, Asian-Pacific countries know Fareed by name, and admire him as much as we do.

A national from the United Arabs Emirates, born in Abu Dhabi, Fareed Abdulla embodies determination and tenacity. He currently is the Senior Vice President for the NEB Asset at Abu Dhabi Company (ADCO) for Onshore Petroleum Operations, and previously served as its Senior Vice President of BAB & GAS Asset and Assistant General Manager of Existing Development.

- He was born in Abu Dhabi, UAE.
- Graduated in Petroleum Engineering, from West Virginia University, 1983.
- Joins ADCO as Drilling Engineer, 1983.
- Reservoir Engineer, ADCO, 1988.
- Head of Reservoir Engineering, Abu Dhabi Marine (ADMA) 1995.
- Fields Development Manager, ADCO, 1999.
- Petroleum Development Manager, ADCO, 2000.
- Senior Vice President of BAB & GAS Asset, 2011.
- Assistant General Manager of Existing Development, ADCO, 2007.
- SPE (Society of Petroleum Engineers) Regional Director for Middle East, North Africa and India, 2012–2015.
- ADIPEC Technical Program Chairman: 2008, 2010, 2013—to date.
- Senior Vice President for the NEB Asset at ADCO, 2013—to date.

Determination in Action

The Choice of Going to the USA

Fareed, tell us about you, about your career.

I was born in Abu Dhabi in September of 1958. I attended high school in Dubai, where part of my family had established earlier. I finished my high school in 1977, and was sent to the USA, for study and obtained my bachelor degree in petroleum engineering.

What was your motivation to study in the USA?

It is an interesting story. The first UAE University was established in AL Ain City in 1976. In the 60s and 70s, we did not have a university in Abu Dhabi, so many high school graduates had to go to other countries to pursue their university studies, especially for majoring in engineering and science. The government would sponsor this through scholarships, and we were given the choice of many Arab countries, Europe or USA. I had planned to study in Cairo, Egypt, completing the paperwork and all preparations to go there, as almost all my friends and classmates were planning to do.

"You Will Go to America"

I informed my mother I was going to go to Egypt for my petroleum engineering degree. But to my surprise, she objected, and announced to me with a very calm poised demeanor: "You will go to America". What a shock!

My mother was a very astute person, with an extraordinary intuition and intelligence, who did not study beyond Quran learning school, but who amazed all of us, leading many family decisions through the years.

I asked "Why America, mother? It is so far from us, do you realize how far away I will be?" But there was no turning back on her decision. She calmly replied to every one of my objections: "For two reasons: you will learn a new culture, and a new language". I was truly in shock, and thought to myself, "she must be joking". But she was not joking at all, and I had no choice. She spoke to her cousin, who was working in the Ministry of Education, and my destination was changed to the USA.

Soon, I landed in Toledo, Ohio, for my English training, prior to my Petroleum Engineering bachelor program at West Virginia University.

How was your adaptation?

Adaptation. What a word for the complete struggle and change of life of those years! It is quite interesting to think about this in retrospect. Unlike our kids today in the UAE, who speak perfect English before Arabic, I basically went to the USA without any knowledge of English. Not even one word, just basic grammar.

Snow and English in Ohio

Someone from Dubai or Abu Dhabi has seldom seen the rain in his or her country. So, when I arrived, in late fall semester, and almost immediately it started to snow, my initial reaction was an anguished "I do not want to be here!" I knew I could go to Kuwait, or to Cairo. And again, at the time I could not understand my mother's decision.

I had to cope with the culture. I could not even ask for food at the diners or the university cafeteria. I decided not to keep liaising with my roommates, who were also from UAE, as I noticed my English was not improving. Because we were finding comfort in conversing in Arabic. We were actually pampered in Arabic: we were receiving the Arabic newspapers from our countries by our embassies. After only two months, I told them: "Guys, I am going to leave you!" And I asked the university student office to assign me to an American family.

The American Mother and Father

I was assigned to the most wonderful couple; their children had already left for their college studies outside Toledo. They became my American mother and father. For me it was a wonderful opportunity to see the genuine American style of living and to understand their culture.

I would join them for dinner, trying to understand and speak English. After a few evenings and corresponding dinners, my American father asked me "What are you reading? What is that small book you always carry around?" It was my English-Arabic dictionary. He told me I was not allowed to come to the dinner table with my dictionary any more. He said that I should make an effort to communicate directly in English, and if I could not understand a sentence from them, they would rephrase it for me until I understood the sentence.

How difficult! Knowing how far apart Arabic and English languages are, this request was surely not easy at all.

But it worked fine! My English–Arabic dictionary was soon history, one relic of my adaptation crutches that I did not need anymore. The strategy adopted by my host family worked so well that I passed the TOEFL examination in three months, and was ready to enroll in Toledo University and later West Virginia University to study my dream career.

Were you alone in the USA?

I was the first member of my family to go to the States, but all my younger siblings came to USA. We are 9 siblings: five brothers and 4 sisters. I am the number four. Big family. A beautiful family!

Love at First Sight

How did you become interested in becoming a Petroleum Engineer?

It was an immediate fascination, like love at first sight. My late father asked me when I was still in high school to accompany him to work. He was in Dubai Petroleum Company, in the distribution segment of the business.

When I saw all the oil barrels, the storage tanks, I wanted to get in the world of Petroleum Engineering. I was determined. I wanted to be a petroleum engineer. Nothing else!!

And in the petroleum engineering world, what is that you were pursuing or liked the most?

This is a question I have no hesitation in answering: Reservoir engineering! While I was at West Virginia, I had to take all mandatory courses for the major, but something that enchanted me was how to prepare a field development plan.

My final dissertation was a field development plan for project that we prepared in a team effort, during the whole final semester of the program. That subject for me was an eye-opener. In my perspective at the time, and still, today, I considered reservoir engineering was the core of our business, the very heart of the oil industry.

The Drilling Years

After graduation, in May 1983, I joined Abu Dhabi Company for Onshore Petroleum Operations Ltd. (ADCO), a subsidiary of ADNOC, the national oil company of my country, starting my work in August.

I wanted to be a reservoir engineer from the beginning, but my first supervisor told me that to become a good reservoir engineer, first I had to be a good Petroleum Engineer. I acknowledged his recommendation and (reluctantly!) went to my assignment in onshore operations, to take care of drilling. My responsibility was to take care of workover and development wells operations. This went on for almost a year. Then, I moved to the processing facilities working in the operations workflows for another year.

After these first two years of fresh-graduate professional field experience, I was brought to the office, to the Petroleum Engineering department, working for two years following Drilling activities, collecting and liaising with the requests from Petroleum and Reservoir Engineering.

Reservoir Engineering Was the Focus

You were still in Petroleum Engineering tasks.

Yes. And I could not wait more, so after 4 years, I went to my boss: "it has been my dream to be exposed to reservoir engineering", I said to him. He agreed to send me to that department for one year, with the promise that after that year I would be back. A year passed by. I went back: "Extend my assignment, please". He did, but only after making me sign an agreement that during or after that assignment I would not ask for a promotion.

Another year in Reservoir Engineer passed by, and I did not ever again go back to that supervisor or sign any other restrictive kinds of papers. I was finally and gladly working in reservoir engineering tasks, preparing field development plans, taking care of simulation of reservoirs, preparing models, arranging scenarios for future forecasts. A dream!

The Shift in the Career

In 1995, I was asked to join one of our Shareholders. ADCO had a program for highflyers, and they offered attachment programs for a year or two with Total in Paris, or at Pau, in the south of France. Alternatively, one could choose to go to the BP North Sea operations in Aberdeen, or at their London offices. Or we had the choice of being attached to Shell in The Hague, or to ExxonMobil in Houston or in Dallas.

Which one did you pick?

I decided to opt for Mobil (before merger) in Dallas, or Shell in the Netherlands for drilling activities in North Sea. It was immediately after my father passed away, and I took it as an opportunity to advance my career. Dallas is a beautiful city, and Mobil had many projects related to reservoir and petroleum engineering. In comparison, the Shell assignment was in a small village in the Holland operations where drilling were handled, which was not attractive to me. So, it was an easy decision for me, I picked Dallas. It was May 1995.

The First Managerial Role

The Mobil assignment was planned to last two years, but in Oct 1995, the management asked me to come back, to fill an opening in ADMA (the offshore Operations Co.), as Head of Reservoir Engineering. It turned out that role was occupied by our shareholders, and instead the management needed and wanted to replace him with a national person. "We know this guy who always wants to be a reservoir engineer," my management in ADCO said. I like to joke about this decision, as I think that what they really wanted is to get rid of me! Having asked so many times to achieve what I wanted, which was a reservoir engineering post, there was finally a clear opportunity related to that, and one way or the other, I obtained what I wanted.

In my new role in ADMA, I had the opportunity to support, sustain and develop offshore ADMA fields.

Then, in 1999, I returned to ADCO, my original company, to what was to be an agreed-upon 4 year assignment. I came back as Field Development manager, taking care of onshore fields. In February 2000, life has its ways, and I became the Petroleum Development Manager. I was in charge of all fields onshore, for the largest operator in Abu Dhabi, keeping my role until 2007, for 7 years. During this period, production output was increased by 30%.

In 2007, I was promoted to Assistant General Manager of ADCO, my first truly executive role.

An Early Realization of Leadership

You are an outstanding leader. I am saying this based on the comments received from your peers in several SPE assignments, from Hosnia, as well as the excellent feedback from some of your team leaders, whom I know. Leadership could be difficult or easy to develop. When did you realize you had a strong leadership?

I have been very critical about my leadership, and have always asked myself if I am doing it right, if my teams appreciate what I do. But in reference to your question, I would say it all started in high school. I always wondered why in high school I was selected for the responsible person to "control" the class. When the teachers left temporarily for any reason, I was called up front to take charge. Why? I am not sure. I have even come to think that maybe because I am tall? Maybe they thought people would listen to me because I was tall. But I truly discovered my leadership through sports. Let me tell you how.

If You Are the Captain …You Cannot Act Badly!

I like very much football, the sport called soccer in the USA, and I have played it a lot. In football, I was always the captain of my football teams, ever since I was 12 years of age! And if you are the Captain …you cannot act badly. People see you. Being a captain is a responsibility. And from there on, I was always held leadership roles.

The football seems to have played an important role in your life. Tell me more, did you play during your study years in the USA?

Absolutely yes! I founded the 1st national UAE Football team in Toledo, and I shaped the team with some classmates who did not have a clue about how to play. So, I would assign them roles: "you are a goal keeper, you are a front runner", and so on. It was a lot of work, as I needed all to participate in the team, and we had only 2 good players. And you cannot shape a team without participation in tournaments. You need the competition. You really do! At the time, my Kuwaiti classmates had become experts in football, and they had gained a deserved prestige playing against the Toledo University team.

Courage in Soccer

After a few months of practice and friendly games, I told my Kuwaiti friends, "We want to play against you". Of course, they laughed, and refused, because they knew we did not have strong players. I kept asking them, and it took me two months to arrange the first game, but they accepted, and at the first match, they beat us 5-0!! Fair enough.

We kept practicing, and after one month, we asked them again for a match, and we lost this time 3-0 and in a third match, we lost 1-0. They did not mind playing against us.

On the 4th match we beat the Kuwaitis, 2-1, and this was really big news. The Kuwaitis were so upset! And they asked us for a rematch the next Saturday on the pitch. But guess what? For 3 months we did not play them. We enjoyed that great victory that cost us so much practice!

The Soccer Learnings

That was the kind of leadership I developed. One where players became excellent players in the end. In the beginning, half of my team did not know how to stop the ball, and they were there only to complete the number of players required for a football team. But we raised our quality for the game collectively, with practice based on the skills of a couple of good players. They helped me teach all how to play properly and efficiently. That was an experience that showed me determination, tenacity and self-confidence can be taught and shared, in order to elevate performances of each individual. I learned that no one is so bad that they cannot add to the team's success.

When I remember all this, I realize that from very early on, I had the opportunity to develop my team with my compelling and motivating style of leadership.

Magic Wands in Leadership

What do you think is the basis of your success?

I can tell you one of my successes at work is to be respectful and to be humble. Don't change who you are because the seat keeps changing. When you respect your people, your team, they will respect you back.

Also, the way I handle personal issues has been a personal distinction. I do not have a magic wand, but I try to help my employees as much as feasible. I make sure I would never ever turn someone down. When I look back I realize I have helped many people who were not even in my unit. With time, people would say "go and see Fareed". I am blessed by having a role that enables me to help. It is not a favor, is part of your duty. It's a great opportunity to support people. As I help people, I have never been turned down when I have been in a situation to request others to go the extra mile. Never.

Of all your achievements and challenges, which one was important for you?

I want to be myself answering this question. I will not characterize the things I do as achievements. Everything I do at work I see as my duty. I am in charge of projects, and they are my responsibility. And that is work.

I volunteer extensively for the Petroleum Institute in Abu Dhabi, and for the Society of Petroleum Engineers (SPE), following Hosnia's shoes. Every other Thursday, I try to go to UAE's SPE offices, and I do it to push and support them to keep growing and to keep the focus in helping the young professionals in our industry. I feel that I have to do this, and there is nothing to brag about my volunteerism.

ADIPEC to the Top

But there must be something that makes you feel especially proud.

To be specific, something that made me quite happy occurred around the time when Dubai was announced the winner of EXPO 2020. I was in charge of the technical program for ADIPEC, and I stood in front of the tech committee, composed by about 100 professional people, and made a declaration. I stood on the stage, and I said, "Excellent achievement by Dubai. They secured the EXPO 2020. Let's draw inspiration from this, and let's make of ADIPEC the most important event in the world by 2020."

Today, I can proudly say that the 2016 edition of ADIPEC already became the largest conference of the oil sector in the world, by number of attendees, highest number of abstracts submitted, and many other key performance indicators such as exhibition space, and gathering industry's key people, etc.

We put Abu Dhabi and the region, on the map, competing with Offshore Technology Conference (OTC), a conference with 50 years of history in Houston, and Stavanger Offshore Northern Seas (ONS). We became double the size of ONS, and we dethroned OTC. We welcomed 97,000 visitors, 10,000 delegates and 700+ speakers in 4 days.

For me that was something special. I know that I shared a vision, and I was part of making it a reality. Today, we can claim that as success.

The Cutting of Green Leaves

Fareed, it seems to me that you had an almost flawless story of success. What were your most important challenges? Those moments, from which you learned life at work is tough?

To remain honest with you, of course, you go through ups and downs. In the downs, the most marked difficulty for me has been when you see organizations shrink and you lose your people. After you remove the "dead wood", and you realize that you still have to cut the green leaves, that really hurts. To cut off those people about whom you care. When you cut the income of a family, is simply horrible. It's a horrible task to do.

You try to convince yourself that at the end, that was part of my job, to terminate jobs, but it is a difficult challenge. And depending on your personality, you will definitively be affected.

A recent big challenge I can think of is the current situation of cost optimization we are undergoing, for which I am the Champion. We had six months to cut expenses, due to the lowered oil price, and I managed to achieve the target of reduction one month before the deadline. I appreciate the level of interaction. We conducted 35 workshops, with a team of only 9 people in a campaign in every department, selling the idea of cutting costs, with an approach of bottom-up. We achieved 200 million dollars of reduction in 6 months, without jeopardizing safety, asset integrity, by relying on improving efficiency and practices.

Am I a Role Model?!

There is a key moment in the life of all leaders when they realize they, themselves, are role models to others. When did this happen to you and how?

It was only a few years ago! It is just recently that I started to hear constantly from the young professionals that they want to follow my path. They say, "You are our role model", "do you need any help in ADIPEC?", or "We want to learn how you do it". Initially, it was kind of funny for me, as I thought "poor guys! I feel sorry for them!!! Who am I to be their role model?!?" But then, I started to value the importance of leading by example, and I began accepting I am facing a mature stage of my career, from which others may gain insight and experience. I am a role model.

Everything I have done, I have done it from my heart, and I am glad the young generations may find in what I do or have done some basis from where to learn and build their own careers.

The Power of Reading

Do you have any personal trait, value, or conviction that you want to share as grounds of your success?

I always basically believe that for me to progress, I have to read, learn, and practice in order to update myself.

I will always be asking how to do things, requesting clarification. In the old days you had to read technical journals. I knew the world of IT in the 80s by reading about it. In 1985, only after 2 years of joining ADCO, I wanted to establish the Computer Club of ADCO. With this aim, I urged my management to purchase 8 computers and with the help of 3 other colleagues, we taught courses after work hours for free, only requesting a very minimal fee of enrollment in our "club". I personally taught *Introduction to DOS, ACCESS, SYMPHONY (Excel), graphics and more.* We created an innovative initiative that was very popular: ADCO employees would book 3 months in advance for our courses that were completely free.

I did this because I wanted to share my knowledge. I suggest everyone get updated, and to read. If you are into management, read about leadership. I try to read everything I can.

Do you have any recommendations for other leaders in their formative years, or perhaps in their first leadership role?

I would like to tell them that once they are in a leadership role, to pay attention to what they say. It's important to carefully watch what they say. Now that you are the boss, anything you say will be considered a fact, a sure thing. If you promise you will "look into your issue", expect that employee to count on you to solve the issue at stake. Everything you say has weight when you lead. People will cherish, analyze, and scrutinize every word you say. Be careful, and at the same time committed. Mean what you say.

What are the challenges the energy industry will face in the future?

I am not worried about the current oil prices, I think market will adjust itself as we go along. The biggest challenge is instead that people have a negative image of the oil and gas industry. Our biggest challenge will be how do we keep an attractive image of our industry so that people are attracted to it? So that we may hire the adequate people to run the industry? How do we capture the knowledge?

The Word Loyalty

What message would you like to send to the young generation?

I think the new generation is unfamiliar with the word "Loyalty." During my time, the oil and gas was the only sector you wanted to join. But today, other sectors have grown very strong in the UAE, like the military, banking, tourism, hospitality, providing many alternative choices. Private sectors are hiring significantly. And if our young professionals do not like working conditions with us, like, for example, an archaic IT system or no café in the company, it's for those reasons that they leave!

Their tolerance and resilience level are low. Very minimal. In our industry we need continuity and expertise. How can you have continuity? With loyalty, with resilience, and with patience.

If we, ADNOC, ARAMCO, KOC, if we all do not become more attractive…. we will not appeal to the young people. They care about the technology. If they find in us to be obsolete with an old style, they will leave.

A Shared Selfie

- **Your favorite word:** Respect.
- **A city:** my home city, Abu Dhabi. And then, Sydney, Australia.

- **An important person of your preference:** After my mother, there are three historical figures that I admire who changed their countries: Sheikh Zayed Al Nahyan, Mahatma Gandhi and Nelson Mandela.
- **A personal happy moment:** Two happy moments: the birth of my first child. Although I was young, and as a young man you do not have the feeling of fatherhood. That realization of "we are three now". We named him Suhail, which means shining referring the second star that you see in the sky at night. It was an incredible happy moment for me. The other, was when one of my younger brothers after chemotherapy and bone marrow was declared free of cancer. I rejoiced with him in a moment of true happiness.
- **Your role model:** My mother
- **A favorite landscape:** Yellowstone National Park, in Wyoming. I have driven several times to the Grand Tetons. I have that image stamped in my eyes.
- **Your favorite food:** I love fish. Followed by seafood.
- **Your favorite color:** Yellow.
- **Your favorite song**: I love music. Cat Stevens (Yousef Islam) as individual singer. As groups, I like Pink Floyd, The Eagles and Supertramp. I guess these selections denote my age!
- **Who supported you the most in your life:** My family. All along!

Post-scriptum

The mention of music is a complex theme for me in the Middle East. Some individuals do not like music, and is an argument to be avoided with them. But with Fareed, it was a pleasant discovery! He was so eager about the question, and very open to share the exact style of music he preferred.

As an incredible flashback to me, when he mentioned Supertramp. It was a favorite of my years during my youth in Venezuela, when I would ask travelers to the US to bring me the LPs, and I could not believe anyone else in the Middle East would know that group. I started singing *"Dreamer"* first verses, just to hear Fareed singing along with me the whole initial verses of the song. We were both singing with the enthusiasm of teenagers.

It was an awesome moment, during one of the very best of all interviews. So fresh and natural, so spontaneous and glad. Fareed is one of the best dreamers I know, and he has been able to convert his dreams in tangible realities and successes for Abu Dhabi.

Dr. Giuseppe Giannetto Pace

"Power is ephemeral".

A Glimpse

Dr. Giuseppe Giannetto Pace is a resilient leader who does not disappoint in achieving the objectives of the organizations he leads. He is still called "Rector" by all, acknowledging his role as president rector of the most prominent public Venezuelan university, Universidad Central de Venezuela (UCV).

With 62,500 students, and 8,600 faculty, the UCV is organized into 11 Schools, which cluster 41 departments, offering 67 degrees, 109 Master, 222 specializations, and 40 PhD degrees. The top-ranked and most revered educational institution in the country, it is one of the oldest universities in the Western hemisphere, founded in 1767. The UCV has graduated more than

150,000 professionals, and is one of the institutions in Venezuela that is beloved and respected by the majority of the political, economic and social sectors in the country.

To lead this centuries-old institution is an exercise of harmonizing live forces that go beyond academics, literally shaping Venezuela's future. To master this kind of leadership is exactly what Giannetto accomplished, overcoming with an incredible resilience one of the stormiest moments in the UCV history.

- 1976, B.S. Chemistry, Universidad Central de Venezuela.
- 1985, Doctor in Chemistry, Universite de Poitiers, France.
- 1994, Department Head of Petroleum Engineering School.
- 1996–2000, Vice-Rector—Academic Affairs, Universidad Central de Venezuela.
- 2000–2004, Rector, Universidad Central de Venezuela.
- 2005, Director, Alumni Association, Universidad Central de Venezuela.
- 2000–2004, President of the Venezuelan Association of rectors of Universities.
- 2006, Executive Vice-President—Education and Commercial sectors, Banco del Caroni.

A Personal Snapshot

Dr. Giuseppe Giannetto is the kind of approachable leader who quickly earns a place in people's hearts. Not only because of the admiration a leader typically fosters in his close associates, due to his high profile, and the importance of his accomplishments. But, in his case, also because he does care about people. Giannetto makes sure that all of his team members, employees or liaisons, feel that they are part of his team. He guarantees that they know the strategy and objectives to pursue, with his excellent communication skills, ensuring optimal workflows and maximum ethics in all his activities. Most importantly, he cares that the support at work and in life of what he calls "his people" is readily available.

Maria started to liaise with Dr. Giannetto in 1999, when she was Reservoir Manager of Petro-UCV, a pioneer initiative of the National Oil Company of Venezuela (PDVSA), to partner with the most prominent Venezuelan university, Universidad Central de Venezuela (UCV), to operate an oil and gas field, sharing the profits. This pioneering idea was to use the joint venture as a mechanism to provide faculty and students with the opportunity of improving oil-related careers with a hands-on approach, inclusive of real data and operational experience. As a true pioneer, Giannetto was one of the main promoters of this unique initiative, and his endorsement was key for many people during turbulent times in Venezuela.

Later on, the launching of the UCV's alumni association enabled Maria to work with Dr. Giannetto again. "Egresados–UCV", as the alumni association was called, was one of Maria's most cherished experiences in leadership. It is one that continues to benefit the alumni of the UCV, the Venezuela's *"House that Defeats the Shadow"*, as its motto says, in Spanish *"La casa que vence la sombra"*.

Arranging the Interview

Meeting with colleagues, with opportunities to remember past anecdotes, is usually time well spent. And to arrange the interview with Dr. Giannetto was for us a welcome opportunity to reestablish contact with an admired professional. He appreciated the opportunity to be part of our project to capture learnings in leadership and resilience. He was especially willing to share his story, as our compilation is aimed for the young generations, a leitmotif of his life.

Our compilation is about leadership and resilience. We already knew before interviewing him that the most valuable lessons we would derive from analyzing his journey towards success were going to be the ones associated with resilience.

Few other people have endured the ordeal he had to go through to defend his role.

From Burgio to Caracas

Tell us about your origins, and how you started your career.

I am a son of Sicilian immigrants in Venezuela. My father came to Venezuela in 1953, leaving his wife and 3 children in a town of Sicily, Italy, called Burgio, in the province of Agrigento. After four years, in July of 1957, having saved some money, he called for his family to join him in Caracas. I arrived to this country at five years of age, and as I have always had facility to communicate, my parents told me I picked up the Spanish very rapidly. I tropicalized!

Where did you live and study?

We lived in the center of Caracas, in the foundational neighborhoods, near the historical sites that shaped the old colonial blocks. The addresses are described by mentioning the two corners of the street, so we lived "*de Truco a Cardones*", half a block away from the main Baralt Avenue.

Oh! In the heart of the city. What do you recall from those years?

Not much, as I was a small child, but of my memories, one of the most intense are the bombs and gunfire of the civil revolt endorsed by the military that deposed the Venezuelan dictator of the 50s, General Marcos Pérez Jiménez.

What did your parents do for a living?

My family did not have economical means, but there was certainly lots of love. Father had a small grocery store and then a butcher store, and my mother was a seamstress. My father had a golden heart, and having been incarcerated during WWII as a prisoner of war (POW), he would help his Italians friends or acquaintances from Sicily in financial need. As a result, we never had an excess of money. He had a butcher shop located "de Amadores a Desbarrancado", which did not do very well, and when it did have good sales, money went to help so many people, that my mother ended being the breadwinner of the family most of the time. I learned social solidarity from my infancy and teen years, directly witnessing how collective support needs to be activated and can effectively help those in need. It was a lesson for life.

The Privilege of Studying

What other memories do you have from your early years?

My recollections are mostly from my studies. I studied in public schools. In the National School called "5th of July", and in the "Agustin Aveledo" and "Rafael Urdaneta", in La Pastora and in San Jose, Caracas. My father wanted to have a son with a university degree, what traditional Sicilians used to call a "*Dottore*", a doctor, in Italian. Anyone who had a university degree, an engineer, an accountant, or a medicine school degree would be called a doctor. And I do not know why I was the one my father selected to be the first "*Dottore*" of the Giannetto family.

My sister and brothers did not attend the university. They went directly to work to financially support the family. I had the privilege to study.

The time came for the selection process to enter in the university, and I was admitted to study in the most prestigious Venezuelan university, the Universidad Central de Venezuela. The news was celebrated not only by me and my inner family circle, but my extended family and all of the neighborhood, as this was received as a victory for all, a collective sense of triumph. I had not even started, but we all celebrated my acceptance.

Chemical Engineering

What career did you choose to study and why?

I did not want to study Medicine, as it was the wish of my father. What I really wanted was to study philosophy. But that is not a career that would have enabled strong earnings. At least, this was my perception. So, a high school professor motivated me, or should I say pushed me, to study Chemistry.

During the studies, I aimed to be first in class. This helped me later on, for the credentials competition to be hired as Instructor Professor. I was chosen among many applicants, in what was a truly difficult journey, of a full day examinations by different technical selection committees.

Another big party in the neighborhood, I guess!

It sure was a great celebration. And worth it, as it was my first difficult journey towards achieving a leadership role. A professor is a leader for knowledge acquisition, knowledge sharing and transfer. I was the first university professor of my circle, and we all celebrated the accomplishment as a collectively earned joyful moment. Remarkable.

You had assured a stable career ahead of you.

In a certain way, yes. The tenure would come years later. I married Yvanka Kancev. Our daughter Carolina was born. We both had a B.S. in Chemistry. She worked in INTEVEP, which is the research center of the national oil company, PDVSA. We always wanted to do more, to achieve more, as this is the kind of education we received, not to be conformists.

I wanted to pursue graduate studies, to go beyond being a professor to effectively lead research and position teams of researchers in chemistry. With the financial sponsorship of the Universidad Central de Venezuela, Dr. Mireya Rincon de Goldwasser, a renowned Venezuelan scientists, member of the Venezuelan Academy of Sciences, strongly encouraged me to go to France for my PhD studies.

Doctorate of State

"I do not speak French!" I objected, but she explained that France was one of the best academic frameworks at the time in the area of catalysis, which are so useful for the oil industry. It was a very competitive area, at the peak of the research in chemistry. I applied to the University of Poitiers, and was admitted.

Our first year in France was complicated. Several elements made it difficult for me to start working in my research area, but during my second year, I started working with Dr. Michel Guisnet, and this changed my life.

Thesis of the Year!

Dr. Guisnet instilled self-confidence in me, and made me realize the capacity I had for research, and to lead. First, he guided me through the thesis of third cycle (the equivalent to a Master of Science). Subsequently, he encouraged me to pursue the Doctorate of State (more than a PhD). He obtained all the approvals, and I directly pursued the Doctorate of State, in one step.

During the last years of my program, Dr. Guisnet let me direct the graduate students in the program, and we published numerous technical articles in peer-reviewed journals of wide circulation. I grew immensely professionally, attending international conferences and representing our research team and Dr. Guisnet many times. My thesis was awarded with the national prize of France "Thesis of the Year," awarded by the president of the National Research department of the country.

Decisions

Due to the prize obtained, when I finished, I was interviewed for a job by British Petroleum (BP). I started to work in a post-doctoral program sponsored by BP, aimed to support research for the chemical processes at the Lavera refinery, in Marseille, at the time the most modern refinery in Europe, with a capacity for refining 200,000 barrels of oil per day. They wanted to evaluate the laboratory techniques to produce the best catalyzers for Fluid catalytic cracking (FCC), so important to produce gasoline. I graduated with a specialization and research experience in what was the peak topic for the refining industry worldwide.

That was one of those non-repeatable moments in life. Yvanka and I, with our two children, had been outside Venezuela already for 5 years, and at that time, we wanted to go back. You are aware, Maria Angela, that Venezuela in the eighties was a country of progress, where research produced many results, especially in the oil industry, and also in the academic sector. I thought I could propel research in zeolites in Venezuela. I was energized!

We decided to go back to Venezuela, and help with the progress of the country. It was an opportunity to pay back with our work for the advantage provided by the scholarship received. We planned to build and shape new futures. We were optimistic.

Starting from Zero

After all that effort, and the success attained, you would think you would be welcomed back at work at open arms. But it was not like that.

We started from zero. I was coming from the big leagues, and started all over with no lab and no equipment. But I was energized, and I started to channel all the knowledge, experience and willingness to do research that I had brought with me with specific actions.

It Must Have Been Hard. What Were Those Actions?

I initiated by creating workshops and courses about zeolites. My work in was related to chemical research for conversion of LPG to gasoline and the pertinent development of catalyzers to that end. I was contacted by the research center of PDVSA, Intevep, and eventually we signed MOUs and contracts aimed to launch research about these topics in the UCV.

With this, I equipped the laboratories of the department of Chemical Engineering. I published a book about zeolites in Spanish, which is still used as a reference for the topic in Latin America. It was a compilation of the state of the art in research about zeolites. There was a second edition, with co-authors.

A leadership role I assumed, at the return from my doctorate, was to lead the Petroleum Engineering School, as Department Head. It was somehow traumatic, because I had to reorganize the department, and let people go. It was painful in some cases, but I was following my ethical and moral values. At the same time, I was advancing in my laboratory research work.

Surprisingly, all my research achievements in zeolites, the articles about catalysis, and my book about research in zeolites, did not provide me the notoriety that a serendipitous incident gave me.

What Was that Incident?

Giannetto replies without hesitation with a surprise statement. It was a peak of the interview).

It was …La Mancha Negra!

"La Mancha Negra"—The Black Stain

As a Venezuelan, I immediately had a very strong "flashback", and recall the huge issue that was experienced in Caracas in the late 80s, when an extended "mysterious" oily stain on the main highway to Caracas's airport caused thousands of accidents. Unfortunately, there were also hundreds of deaths, to the point no one wanted to drive to the airport, when the "Mancha Negra" was developed. When it was removed, it would rise again, on the asphalt, puzzling technicians and asphalt experts. It was deadly.

Were You Involved in the Analysis of La Mancha Negra?!

I like to say that I led the solution, more than the analysis. Let me briefly explain.

La Mancha Negra first appeared as a smudge 50 m long, noticed by asphalt road workers, and you might remember the oily substance made the roadway extraordinarily unsafe, causing vehicles to crash, spinning unsafely in all directions. It was a serious problem, claiming lives and without a visible solution on sight.

I was nominated by the UCV as lead researcher for the Commission of the Ministry of Transport, Ciro Zaa. I co-led this research with the Energy and Mines Minister, Mr. Arrieta, and the President of PDVSA, Luis Giusti. I was no expert in asphalt, but this was one of those things of fate. I was in the right place at the right time. The true cause of the black stain was not corruption in asphalt quality control, nor was it bad formulation of the asphalt. In one word, the asphalt was not the issue. The government had brought specialists from Norway, United States, Sweden, Germany, but they had never consulted a chemist. On the field, analyzing the substance, I proved that the black stain was not due to a flawed asphaltic layer. I collaborated with my colleagues from Intevep and the UCV, to reveal that the issue was the old cars and buses leaking oil, a result of a very old auto motor system. As an average, private cars had 12 years of usage time, and buses, 20 years.

Academic Vice-Rectorate of UCV

In the UCV, the top roles are won by election, through a democratic process. After my tenure as Department Head, I was nominated for Vice-Rector—Academic Affairs for the UCV.

I ran, and I won. I won because I was known as an academic, a technical professional, removed from the political games and interests. This new role of Vice-Rector definitively triggered a different period in my career.

I wanted to implement positive changes in the academic standards, across departments and schools in the University. I wanted to hold the role Vice-Rector only for the established four years of the term, to then go back to my academic work.

And How Did You Decide to Run for Rector?

Life has its ways, and in 2000, the strong support of the pro-President Chavez' Party were endorsing another candidate for UCV's President role. The democratic and progressive leaders of the UCV considered that if the pro-government parties would win those elections, the university was in real danger of losing the democratic framework within which it always flourished.

It is difficult to explain, even to myself, as I always had rejected administrative roles. But it is also true that I apply myself to do everything in the best possible way. When I work in anything, I do it not as an obligation, but as a compromise, so I do it with a genuine commitment. I discovered and proved I could lead. My colleagues saw that in me, that capacity and strength. They wanted me for a special circumstance, to face a higher power that was menacing the democratic framework of the UCV.

In this segment of the interview, Dr. Giannetto expressed his inner beliefs about leadership, and a few legitimate pearls came to light, expressed with such simplicity, that I was impressed to see his ethical stature and his humbleness.

And that Is How It Started

A collective petition of numerous sectors of UCV urged me to accept my nomination as Rector and run for the elections. After carefully reflecting about this heavy involvement in what ultimately was to be a more politically-related role, I accepted.

We won, and with a sweep win! Against all polls, that favored the government forces, for many reasons, inclusive of the huge "marketing" campaign, we won. That is how one of the phases that marked my life started.

Power Is Ephemeral

I was convinced that if you have a leadership role, you have to make the decisions. If you do not, no one else will! So, to lead just to leave the status quo you find, is idle, is not ethical.

I had in my head that the roles are just temporary. That power is ephemeral.

You are not a Rector. You are not a Vice-Rector. What you are doing is occupying that role circumstantially. After the term, you will return to your life. One assumes the roles to help, to implement a vision, to lead. Not to use the roles to benefit from them.

Did Your Family Have a Say in Your Decision to Run for UCV Rector?

Thank you for asking this. Yes. It is important to share that all the important decisions in my life have always been consulted with my wife and children. And I wanted to be true to my ethical values.

Ciudad Universitaria de Caracas, the Venue for the UCV, as UNESCO World Heritage

Tell us about your work as UCV Rector.

The first six months of my tenure were a sort of a honeymoon, where all was smooth, happy, and bright. Assuming my role, I continued several initiatives that had been initiated before me. I accelerated the activities that I considered were key for the UCV. A first great achievement was to attain the certification from UNESCO for the "Ciudad Universitaria de Caracas", the venue of the UCV, as World Heritage. It was not a trivial thing as you may imagine. Several elements were needed to achieve this. The UCV had lost its accountability for the Botanical Gardens and the Escuela Tecnica Industrial, two large areas connected to the architectural wonder that Carlos Villanueva, the architect of the UCV had envisioned as integral part of the design.

We were in a quest for the designation, so we had to attain those restitution.

But How Could You Achieve that, If Those Were in Hands of the Government of President Chavez, and You Were Clearly not an Acolyte?

Incredibly as it may seem nowadays, back at the time, Nelson Merentes and Hector Navarro, very pro-Chavez, and currently Ministers in the President Maduro's government, had worked closely with me while I was Vice-Rector. At the time, you would work with the people based on their professionalism, not on their political affiliation.

In 2000, the UCV venue, the "Ciudad Universitaria de Caracas" was declared World Heritage site by the UNESCO. It was an extraordinary achievement!

(In this segment of the interview, I knew what was coming, and was in expectation to see how this important leader would tell me his side of the issue. The events he is about to relate were violent. It is a ironic, that he is the only rector of the UCV in history with a last name that translates as "peace". His full name is Giuseppe Giannetto *Pace*).

The Nonsense

You can imagine the atmosphere during those months. There were celebrations, and all was shining, all was bright and golden. The people had an academic rector, with whom they identified.

But in the government, radical groups were working and planning to take the UCV by assault. As I did not have political background, I did not recognize any signs of what was about to happen in 2001.

On a regular day of session of the UCV Supreme Council, on March 28, 2001, a group of people took the Council Conference Room and the rectorate building by force. There was tear gas thrown on us in industrial quantities, for days in a row. It was madness in action. They were a group of people, radicals, composed by students and employees, who were determined to stay in the rectorate offices, with the intention of installing a new UCV government structure and authority.

At the beginning, we all thought it was an action of some radical students, as there had been actions like that in the past, but those were short-lived, and solved with dialogue. We soon realized this was a different kind of action, specifically aimed to force the Rector to resign. The President would then appoint a temporary rector to "normalize" the situation, gaining the leadership of the UCV, creating an obvious stronghold against the government.

The Crisis

We tried to solve the situation talking with the students. As usual, we wanted to establish bridges of communication. We did not succeed.

How Long Did the Crisis Endured?

From 28 March to 3 May of 2001, a little bit more than a month.

The UCV had become the only democratic institution in the country that resisted the changes of the government of President Chavez, and it was the non-submissive fortress of the democratic forces of the society. But not all in the government endorsed such action. Some pro-government politicians (not radicals) mediated and provided support without which I could not have resisted.

How Did You Resist?

I stayed six days inside my office in the Rectorate, day and night, without going out. Not because I was under a kidnapping, but because my intuition was that if I left, the people who had taken the office, violating the institutionalism of the rectorate and the UCV.

I was left alone with five guards, who remained loyal against all odds. The rest left, and we cannot blame them. It was an extremely violent situation. Sergio Trocelt, one of the guards, informed me "Rector, outside is Professor Agustin Blanco Munoz, who says that if you abandon your office, he personally guarantees your physical integrity". "What should I tell him?" I was alone in my office. After careful thought, I decided to stay, determined not to enable the take-over of the UCV by assaulters. First dead, I thought, rather than putting the Rectorate on a silver platter for them to take. That was not to happen under my tenure.

Well, I told Trocelt a curse word, that I will sanitize it for you: "You can tell Blanco that he can go to where he wants, but I will only leave this office dead".

The media coverage started to become intense, and it was 24/7 coverage. Media coverage was fundamental to resist and maintain the UCV under its proverbial institutionalism and democratic setting. If it was not for the media, we could not have shown that we resisted leaving pacifically, and that the violence was at the other side. It became clear the assaulters wanted to take control of the UCV, in order to put it at the service of only one ideology. Luckily for us, and unfortunately for them, we started a pacifist resistance. The resistance demonstrated the problem was not against me, as Rector, but a planned action of assault that, if successful, would have implemented that modality as a way to take over the institutions not won by electoral results.

In that moment, they also tried to assault other universities in Venezuela, like University of Carabobo, and University of Zulia. But they could not succeed, because the other universities were prepared, alerted by what was occurring in the UCV.

The Most Difficult Moment

The University emerged from this problem reinforced, strengthened. I became a symbol of the University autonomy, as the institution as a block endorsed the figure of the Rector, receiving the support of all sectors of the opposition to the government. And there were even some segments of the government which were not in agreement with such violent actions.

It was a psychological war, besides the physical presence by force in our offices. They called my home. They shot my body guard, and the body guard assigned to my son. Brutal, indeed.

I had to make a plan for my family, and will never forget that the embassies of the USA and France offered support, to pull me and my family out from the country. I told Yvanka, "you will leave with the children, but I will stay, no one will take UCV by assault. Only over my dead body." Once more, I received the support of my wife.

It was the most difficult moment of my career.

I stayed in silence for a while, impacted by hearing for the first time the details of what was an issue televised in real time, that made all realize the government followers were not using democratic means.

What Happened After the Incidents of 2001?

I ended my period peacefully, and after my four years, as planned, I had the pleasure of handing over to a person democratically elected for the role.

A New Phase

Afterwards, and along with a group of other professors, we decided to launch the Alumni Association of the UCV. An institution that graduated 150,000 alumni who did not have such a pool of ideas, of people, of networking opportunities. You were the Executive Director, Maria Angela. You must remember.

I Do! It Was a Fantastic Journey!

Yes, it was an initiative for which we gained the endorsement of the main private organizations in Venezuela. The productive sector contributed to Egresados-UCV, with ideas, funds, people. It was a great journey indeed.

I have come to realize that when you build credibility, there is a strong trust in your counterpart. I have the gift of communicating, and of building trust based on my results. I am convinced that the capacity of communicating a vision and of building trust is an integral part of leadership.

What Are You Currently Dedicated to?

I am now executive advisor of the Grupo Caroni, with a bank, and two private universities in their holdings. I coordinate the actions of the Group for the sectors of Health, Business, and Education.

What Has Taught You More in Your Career?

I have learned more from the crisis than from the happier moments in my career. The punch-downs make your reflect. The knock outs against you and your team make you ask why? The victories do not.

What Message Would You like to Send to the Young Generation?

To be resilient. To persist. There is a poem of Rudyard Kipling, who won the Nobel Prize of Literature, which is called "If", that I would encouraged them to read.

(Here, Giannetto recites a few verses of "If" in Spanish).

"If you can keep your head when all about you
Are losing theirs and blaming it on you,
If you can trust yourself when all men doubt you,
But make allowance for their doubting too;
If you can wait and not be tired by waiting,
Or being lied about, don't deal in lies,
Or being hated, don't give way to hating,
And yet don't look too good, nor talk too wise:

You have to believe in your objectives. The small hurdles and the big ones".

A Shared Selfie

- **Your favorite word:** Perseverance and love. In other words, never give up, to be resilient. And to help, love, care about others.
- **A city:** Rome. It is my favorite city in the whole world.
- **An important person of your preference:** Nelson Mandela. An example for anyone.
- **A personal happy moment:** The birth of my granddaughter Carlota. Everyone around me told me grandchildren were a different kind of love. I did not understand it. Until Carlota arrived to our lives. That was the happiest moment of my life.
- **Your favorite food:** Pasta with meatballs. The Sicilian pasta with meat sauce and meatballs that my mom used to prepare for us.
- **Your favorite color:** Blue.
- **Your favorite song:** "Imagine", by John Lennon, a wonderful call for brotherhood.
- **A landscape to remember:** the Avila, the reigning mountain that overlooks Caracas.
- **Who supported you the most?** Firstly, my faith in God. And the person who supported me the most has always been my beloved wife, Yvanka.

Post-scriptum

Saying farewells, I said a few "arrivederci", an Italian farewell, to Giannetto. He asked me if I knew that originally he did not speak Italian.

I could not believe it! One iconic figure of the Italian community in Venezuela did not speak Italian? Why?, I asked. He explained that at home, his parents spoke only Sicilian. He learned perfectly the Sicilian, but not a word of Italian. Only when he became Vice-Rector he learned Italian, as he became a representative not only of the Sicilians but of all Italians in the country. Italian-Venezuelans would address him with long conversations in Italian, to which he replied in Spanish. Gradually, he learned Italian, Now, as he says, he speaks Italian "with a terrible Sicilian Accent".

Dr. Pinar Oya Yilmaz

"Tenacity was foundational to my success".

A Glimpse

The external face of ExxonMobil in international liaisons is a woman, Pinar Yilmaz. To say that Pinar knows all the main executives of the most important operating and production companies in the world is an understatement. Pinar knows not only the executives, but also the key people who can clear roadblocks for difficult deals or who may add their technical insight, so that decisions are made with a "win–win" approach for all stakeholders involved.

In 1997, ExxonMobil created a role that leveraged Pinar's countless abilities, "External Technology Coordinator," in the Technical Organization of ExxonMobil Exploration Company. At that time, women in the leadership of the Oil and Gas sector were even scarcer than today, where they are still a notoriously small minority. Pinar held that position until 2009 when she became the "External Collaboration Advisor" in the Planning Organization of ExxonMobil Exploration Company. In this role, Pinar is responsible for coordinating external geoscience activities with universities and geoscience research centers globally. Pinar also manages interfaces between professional societies and ExxonMobil for the upstream companies.

Additionally, her volunteering work for the most important professional societies of the Oil and Gas sector has been recognized with numerous awards in several regions, from Brazil and Venezuela, to the United States and more. She is a constantly invited contributor at the steering committees and boards of key organizations, like the World Petroleum Council and The National Science Foundation. Here are a few milestones in Pinar's career:

- M.A. in Geology from Bryn Mawr College, Pennsylvania, USA.
- Ph.D. in Geology from the University of Texas at Austin (1981), Texas, USA.
- Professional interest is focused on the geology and tectonic evolution of the Tethyan system in the Alpine-Himalayan belt.
- Joined Mobil Exploration and Production Services in Dallas in 1980.
- In 1984, she joined Exxon Production Research Company where she served in a variety of technical positions until 1997.
- External Technology Coordinator, technical Organization, ExxonMobil Exploration Company, 1997.
- External Collaboration Advisor, Planning Organization of ExxonMobil Exploration Company, 2009.

A Personal Snapshot

In our eyes, Pinar has been a key contributor to every event of relevance for the industry. With respect to geologic-focused events, we have seen her influencing teams making decisions at a variety of professional societies' events, especially at the annual meetings of AAPG, SPE, GEO, and other professional societies.

Pinar has been a key liaison for the oil industry in Kuwait, and the Middle East, as well as for Brazil and Latin America. We have seen her become a foundational pillar in many of the initiatives for the advancement of the oil sector, directly with the official relationships that ExxonMobil maintains with the national or private operators, or through the programs of the professional societies.

Pinar is a natural leader who leads by example and supports many other leaders in the industry in their own growth journeys. She always provides an insightful recommendation, a strategic vision and a powerful capacity to articulate paths to implementation or solutions.

Arranging the Interview

Pinar was interviewed in Houston, Texas, in the United States. Houston is her home, which is logical, for a woman who has dedicated her professional life to the oil industry. We had a long conversation, where ideas, memories and time flew freely, revealing that other side: a more intimate and personal Pinar. The Pinar who is sensitive to the young generations, the one who cherished her memories deep within her, and who honored us by sharing her insights on life and work.

She was interested in knowing more about what sort of project we were putting together. We explained that we still were not sure if we would compile our findings and conversations on a website or in a book, but that we definitively wanted to capture her story of success and, most importantly, her insights into leadership, so that we could inspire others as she inspired us. She agreed to join our project, which would have been incomplete, if we couldn't have included a person who has supported, mentored and guided so many!

One of Those Things

Tell us how you started in the oil and gas industry.
What a question! Let me start by academics. I studied my Bachelor's Degree in Pennsylvania, at the Bryn Mawr College. Then, I was admitted into the University of Texas at Austin, for my Ph.D. studies. At the time, I was focused on pursuing my Ph.D. to go back to work in Turkey.

During the last year of my Ph.D., I was offered a job by ExxonMobil, at the time Mobil, and all my advisors pushed me to take it. I was not certain that it was my preferred option, as I had other ideas, closer to my love for pure geology, without narrowing my focus on only the petroleum geology side of my career. Plus, accepting the job offer meant I would remain in the US, rather than working in Turkey. But my advisors, and colleagues all guided me and advised me to accept the offer. I considered the benefits, opportunities, and advantages of the offer, in addition to the stellar reputation of the extraordinarily prestigious oil company, ranked first of the world by so many accounts. It was not an easy decision.

I took the job. That is how I started in the Oil and Gas Industry. It's one of those things, that your career starts in one direction, and you take charge to evolve your own way.

The first half of my career was spent in Research and Technology (R&T), then I worked for a couple of years in Production, and then I've worked the duration in Exploration. I guess I have followed the reverse of the path that many other geologists take, who generally start in Exploration, and then migrate to Production, or to R&T. But this was my path. And I loved my job from day one!

I have travelled a lot, even in my initial years in the technical job, when I was not yet representing ExxonMobil on an international scale. I often travelled to attend technical meetings with our partners all over the world as a geologist in charge of specific projects of R&T or Exploration. Sometimes I travelled to fulfill my responsibilities as a member of an organizing committee for a technical event that a professional society would organize in a country outside the US. This exposure to other cultures, other organizations, geologic settings and production schemes, gave me a valuable experience from which I gradually built my own insights and relationships. These are relationships that I keep building, as our industry evolves and changes.

ExxonMobil Created a Position for Me

In 1997 ExxonMobil created a position for me.

Pinar, how is that even possible in ExxonMobil? I would have never imagine a company like ExxonMobil would engage in these sorts of exceptions to the status quo, to the rules, to their organizational established roles!

Yes, and to tell you the truth, I did not imagine this would happen either.

It was the first time that ExxonMobil created such a role for a specific individual, and I really appreciated the gesture. The role they created for me was entitled "*External Technology Coordinator.*" It was a win–win opportunity.

I have been evolving my role since that moment, because the job keeps changing, from internal technical projects to external strategic planning-related projects these days. It gives me the opportunity to steer my progression with a different focus, as the business trends in different directions. I can pursue new technologies that enable production from new areas, strategizing about how to beat challenges posed by low-price of oil, or our company's approach to face political disruptive events in certain areas.

The job has three parts: (1) Universities and projects with the academic sector, following the objectives of Exxon's global business regions. (2) Coordination of activities with the National Oil Companies (NOCs), and Research centers around the world (3) Coordination of sponsorships and collaboration of ExxonMobil with professional societies.

I work for the global upstream organization. I work with teams when they need insights from external sources. And I work with executives when they will work with external professional societies with a global outreach.

Yes, this was the first time ExxonMobil created a role for an individual employee. I appreciate the significance, and I keep enjoying this role every day.

The Most Profound Experience

You have a long list of successes while working in several roles. What experience in all these years impacted you the most and why?

Pinar thinks and pauses to then select a fantastic, field-related example.

I am going to select one example. I want to tell you about how my field geology experience was inspirational. I went to Alaska as a site geologist, and came to the realization that this business is real. You depend on that drilling bit, and the results that come from a mud log, and you make a call with regard to when to stop drilling, and where to perforate.

You are responsible for whether the well will be a producer well, and not a "dry", non-productive well. You are leading the team to decide when the drilling bit has reached the right depth, from where oil is to be produced.

Yes, there is no doubt. I remember being there and thinking: "*This is exactly the role for which I have trained!*" This is not an office job. This is not visualizing tectonics or regional geology models as concepts. No. These are the skills needed, well by well, to secure energy for the world.

Yes! I think that was my most profound experience. It must have been in 1982. It hit me then that drilling wells was the core of our business. There was snow. There were storms. I worked night shifts. It was hard. I worked long hours at the well, checking samples, in freezing temperatures, with this big responsibility. And I was the first woman. But everyone was very aware you were there. I had my own bathroom in the rig. People went out of their way to be nice to me. Amazing.

That experience completely changed my outlook on the industry, on my profession.

Where Your Salary Comes from

I drilled a total of 78 wells while I was in Production. This is when you learn from where your salary comes. There is no doubt. Drilling gave us the oil barrels to sell. When there is no oil, there is no income.

I highly recommend to any geologists that they go to the field, so you see your results. You feel the excitement of your knowledge in action. From safety, to interpretation, to engineering.

Then, you incrementally increase your understanding of the great integration and interwoven workflows of our fantastic industry: drilling companies, the mud loggers, the well loggers, the seismic crews, the fracture jobs, completion jobs, pipeline crews, the oilfield service companies. It all comes together. You have to make many decisions. A lot of decisions. And you, as a geologist, make the decision. You gain so much. I gained a lot.

It was good to have that insight, that knowledge and experience early in my career.

Unfortunately, these days, very few people go out on rigs. I am convinced they do not know what they are missing, because if they did, they would demand field experience. But there is no other way to acquire field experience. You just have to go out there, in the field, wherever that may take you.

"You Will Be Independent. You Are Going to Achieve Results. No Failures"

Some people had a special influence in our careers. Who was this person for you and why?

My mother. She had two daughters: I have a younger sister. My mother raised us with a leitmotif for our life: "*you are going to be independent, you are going to achieve results. No failures.*" That was first. The mentorship, guidance, support, sponsorship, confidence, advocacy of my mother was there for us. She was definitely the most special, important, and best influence in my life.

Then, I had some great Professors, who allowed me to think independently. I owe them a lot.

Finally, in Exxon I had two mentors who taught me so much. I still correspond with them. They had senior leadership roles, and although they are retired now, they have always been there for me. Showing to me, by example, how to face work with success. They taught me about dedication, commitment, discipline, and that it's important to laugh! To enjoy the work.

I Do Not Think I Am a Role Model

Was there a moment when you realized you were a role model?
I do not think I am a role model. All we working women of my age, or similar age, were pioneers. I was privileged, to be in America, with freedom of thought and behavior. I would have never thought I was in any way considered a role model. I still feel that way.

Some other women did not have my luck, and had to struggle, fight and even work harder than all around them, to gain their deserved leadership role in other parts of the world. I count several of those among my closest colleagues and friends.

I have had moments, though, when leaders of substantial importance have revealed to me they were positioning themselves in higher roles, rather than concede it was a posture of defense in response to women rising to incrementally higher roles in the oil industry. There was one very senior curmudgeon, a geologist by education, who told me: "I gave you a good tip once, but it's not valid anymore." He then elaborated that what he had told me was "Do not smile so much". I remembered that! I took it, at the time, as a recommendation to look more professional, but he explained "*we men were threatened, we were not used to have women around*". He conceded the recommendation of "do not smile so much" was a cautionary recommendation aimed more from the perspective of the men, to protect them feeling awkward and not knowing how to treat us females at work. I realized that the portion of his comment "*but not anymore*" meant he had realized his recommendation was for himself, not for us. It demonstrated his insecurity having more women at the workforce. We all had to learn how to deal at work with a growing reality, of more women in the oil industry.

At the Third "No", I Stop

What are your personal strengths that you consider propelled your success?
Tenacity. For me, tenacity was foundational to my success, and it is a strength very much grounded in my family education, in my values. With

time, I had to develop other skills and I would like to think I have well-developed diplomatic and social skills.

In my own field, I like to think that I have very good technical skills. I keep contributing to technical committees, and aim to keep myself attuned to the new developments in geosciences and reservoir studies.

There is another strength that I cherish in my toolbox: Determination. I stand on the grounds that there is nothing I cannot do. I do not take no for an answer. But I have learned, with the succession of bosses that I have had throughout my career, that if I get a "no" response, I would try for two more times. At the third "no", I stop.

I prepare my arsenal of convincing arguments, with more business reasons, more detailed technical reasons, and even diplomatic reasons. They usually say "yes", especially if you are defending the cause of other people, if you have demonstrated you are a loyal employee.

I have never asked for myself; I have always asked on behalf of other people. It is always for other people that I would argue and request something.

The Industry Is Very Different Today

Pinar, in your opinion, what are the challenges the energy industry will face in the future?

For someone who has spent 30-plus years in the industry, I can tell the industry is already very different today than it was before. We will be very efficient, and aligned internally. We will be dealing not only with business considerations, that is, ensuring economic profitability, but also how to work in entirely new ways.

And we will have to find ways to integrate the new generations who grow up with technology-based interactions using smart devices. They do not engage in face-to-face interactions, but instead prefer to work remotely, and individually, using what is only in appearance an integrated approach. Each one engages in a piece of the solution. Only a partial piece. So, in the future, we will have to figure out how to blend the new generations into teams that are required to provide integrated solutions. If the young people do not work together, we will lose the best of our integrated, multidisciplinary approach to exploration and production. This is the main challenge I see: how can we blend them into teams that the future industry needs for integrated work.

The World Is Larger Than Your Inner Circle

What message would you like to send to the young generation?

Only one message: work hard. Work on things that you actually enjoy, for the benefit of the greater population, in order to contribute to something greater than yourself. I would advise them not to be too selfish.

There are so many ways that young people can contribute in the energy industry, not only in oil and gas exploration and production. I would definitively tell them they will have to work hard, and not to have an attitude of entitlement. Realize that the world is larger than your inner circle.

There are a lot nowadays in the leadership who talk about resilience. What importance do you assign to be resilient?

I do not think I have thought about it as a factor in itself. But I think we all had to be resilient to survive!! (*and at this point, for the first time in the meeting, she laughs loud and long*). We had to bend, we had to change, adapt, as our jobs changed. As our environment changed. If you are not resilient, you won't make it. You may be in an office or in a mountain cabin, with your own self-made tools to survive. You have to be tough! For me, that is an integral part of all of us.

I think about resilience as an intrinsic element related to my Turkish, very traditional, very secular family. An example is my coming to America. I have been living here since 1972, with no family members, except those whom I made my self-selected family. If I was not resilient, I would have gone back to Turkey immediately, in less than a year.

Your self-selected family?

Yes, I have family in America. When I first arrived in America, their daughter was going to the same university as I attended. I visited them for ten days, and ever since I have considered them my family. They are a dear group of people for me, my selected family, who over the years have shared so much with me. Our families have become close. Their daughter is like my sister. She works in Medfield, near Boston. She is an Interior Decorator.

A Shared Selfie

- **Your favorite word:** Happiness.
- **A city:** Izmir, in Turkey. And my second favorite city is Florence, Italy.
- **An important person of your preference:** My mother and my two nieces: Betul, an architect who works in Moscow, and Begum, who works as a Chinese guide in Turkey.

- **What do you like for spending your leisure time::** Exotic vacations, going to the beach. Nature. I love to read. And above all, I love opera.
- **A personal happy moment:** I am happy about very simple things. Blooming flowers make me happy.
- **Your favorite food:** Artichokes.
- **Your favorite color:** Turquoise blue.
- **Who supported you the most?** My mother made a lot of sacrifices for my sister and me. My sister is an O&B Surgeon. I am the rock doctor and my sister is the human doctor, the real one! People sometimes confuses us, and asks me about health issues, and I have to tell them "you have the wrong doctor!". So many people, I cannot even list them, have supported me.
- **A death:** My mother passing away, seven years ago.

Post-scriptum

The volunteering work of Pinar in the professional societies of the oil and gas sector is impressive, and it is worth mentioning it here, to complete our effort in profiling our dear friend Pinar.

- Member of the American Association of Petroleum Geologists (AAPG), the Society of Exploration Geophysicists (SEG), Geological Society of America (GSA), European Association of Geologists & Engineers (EAGE), Houston Geological Society (HGS), Society of Petroleum Engineers (SPE), and the Geological Societies of London, Brazil, Turkey, and Nigeria. She is a member of the World Petroleum Council's Executive Committee. Pinar also sits on the Cambridge University CASP Consortium Scientific Advisory Board and on advisory boards for the Women's Global Energy Leadership Conference and the National Science Foundation's project, 'On the Cutting Edge'.
- Relevant and recent roles in professional society committees:
  - Vice President Finance of the World Petroleum Council (2014–2017).
  - Member, World Petroleum Congress 2014 Moscow Executive Committee.
  - Member, International Petroleum Technology Conference (IPTC) program committee 2016 Bangkok.
  - Technical Program Committee GEO Conferences in Bahrain 2000–2016.

Lionel Levha

"There is always a solution".

A Glimpse

Lionel has received the highest recognition of France: He is a Knight of the Legion of Honor. The award requires the flawless performance of one's trade, as well as doing more than ordinarily expected, such as being creative, zealous and contributing to the growth and well-being of others. For the French, the

Legion of Honor is the distinction of the highest order. This honor with worldwide prestige was awarded to Lionel Levha, one of the most prolific executives of TOTAL, the French oil and gas giant.

Lionel is very deserving of the distinction; besides having an extraordinary career in a variety of roles in different countries and regions of the world for Total, many of the main technical conferences organized for the Oil and Gas sector in the Middle East during 2011–2016 include his name as Chair, Co-Chair, Sponsor or Advisor, for the Society of Petroleum Engineers (SPE), the European Association of Geosciences and Engineering (EAGE), and the Society of Exploration Geophysicists (SEG).

With engineering degrees in Energy and Petroleum Engineering from the French Petroleum Institute School, Lionel Levha joined Total in 1981. He has succeeded in accumulating numerous achievements in more than 35 years of experience worldwide. In 2016, Lionel was appointed Vice-President, Total EP Al-Shaheen. As such, he leads Total's activities in this Joint Venture JV with Qatar Petroleum, to operate Qatar's largest offshore oilfield, with a gross production rate of 300,000 barrels of oil per day.

- Engineering degrees in Energy and Petroleum Engineering from the French Petroleum Institute School.
- 1981 started at Elf, which later merged with Total.
- 1987 Deputy Operations Manager, Elf The Netherlands.
- 1989 Exploration Manager, Elf Angola.
- 1991, Joint-Venture Manager in Elf UK, London/Aberdeen.
- 1995, Vice-President, Elf Affiliates CIS (Russia, Azerbaijan, Kazakhstan, and Turkmenistan), Eastern Europe countries and China.
- 2000 Executive Manager, Dolphin Project, UAE and Qatar.
- 2006 General Manager of Total Italia.
- 2009 Managing Director of Total EP Madagascar.
- 2011 General Manager of Total Kuwait.
- 2016 Vice-President Al-Shaheen, Total EP Qatar.

A Personal Snapshot

For us, Lionel was a key figure in the launching of pioneering initiatives with the Society of Petroleum Engineers (SPE) in the early 2010s. We were working in several executive committees organizing a series of technical conferences, as chairs, co-chairs and coordinators. These initiatives were new conferences for Kuwait and in the region with the SPE. For example, we led the Program and Executive committees of first-ever events in the region and Kuwait, like the Heavy Oil Conference and Exhibition (HOCE), and then the Kuwait Oil and Gas Show (KOGS), the flagship event of Kuwait for Oil and Gas sector. Hosnia was the Chair, Lionel was the Co-Chair and Maria was the Coordinator of these conferences.

Facing and overcoming the challenges faced in the organization of these events was not an easy task, as at the time (2013–2016), there were no adequate convention centers in Kuwait, like there are today. We benefitted greatly from the great support and leadership of the Kuwait senior management in KPC and the so-called K-Companies (Subsidiaries of KPC) who steered the efforts, guided by provided guidelines. But it was a result of the attention to the everyday tasks, in great part done by Lionel and us, that these events were a great success from the first offering. Ours was a winning combination that produced many fruits for the expansion and success of SPE in Kuwait.

Present at all times, with a perspective based on his worldwide experience, Lionel provided a strategic vision and an energetic capacity to lead and accommodate all visions in a consensual path. This is the only way things work, not only in Kuwait but everywhere else. Lionel's calm but firm management style with the professional societies of our sector was one we truly enjoyed.

Arranging the Interview

We contacted Lionel to interview him while he was still settling into his new office in Doha, Qatar. Open and very accessible, as is his style, he enthusiastically wanted to contribute to our project focusing leadership and resilience. We were glad to record his learnings and perspective.

Lionel had many learnings to share, and it was a privilege to hear some of the anecdotes of his professional life, which has taken him many places. His love for excellence at work became evident from the beginning, but most importantly, we were fascinated by the importance he assigns to people. We observe that leaders reach pinnacle roles not because of technical deep knowledge, which is of course important, but primarily because they care about people, and about their teams' members.

We conducted the interview after working hours, and even if it would have been understandable that he would be a little bit tired, the gleam in his eyes when referring to the most difficult or important moments in his career was there, as if it was early morning time. He re-lived those moments for us with the passion, as it was the first time he was telling those stories. It was an interview to remember.

The Start of the Journey

Tell Us How You Started in the Oil and Gas Industry. Tell Us About Your Career

It's a long time ago! It is almost 36 years now. Everything started in the Engineering School. I studied in the French system, and in this system, I studied a Master, which required a 3 months training, that I had to perform in a governmental or private organization related to my field of study.

This first industrial training of my studies for me was organized at Elf, a company that merged into what is today Total. At the time, that was the training in my second year of engineering, just before the Master, at the *Ecole* of Engineering. I engaged in this training, and prepared a final report. At that moment, I was not just fulfilling my student duty, but discovering what was to be my life: the oil industry! My supervisors at Elf liked my work as a student. They called to see if I would like to do my long-internship with them, which was scheduled to occur during my third year.

And I accepted that offer.

North of Timbuktu

In France, at that time, I had to perform the military service. It was mandatory for all men of my age, but you had the option of doing a longer industrial service that would qualify for the governmental requirements, to count as a military service. Mali, in western Africa, fit that purpose, so I was sent to Mali, in Africa.

Africa! for How Long?

I took this opportunity, extending the original one-year assignment to 18 months, so that it could count towards the military service. I had everything to gain from this opportunity, as I would receive professional training, and be directly involved in engineering activities. I would work in activities of the company, fulfilling a dual purpose: engineering and service to the communities. I performed this service with Elf, the company that, in time, would be merged and then transformed into the French oil and gas company, Total.

After my training, Elf hired me immediately, even though I had not yet finalized my engineering studies. I obtained a special delegation, like an official permission, from the engineering school, that enabled me to start working, and that is how I started drilling wells in the North area of Timbuktu.

Did You Receive Training at Your University About Drilling Wells?

No, I was originally pursuing my degree in Thermal Engineering and then in Petroleum Reservoir Engineering.

Tell Me More About This; It Is Remarkable, to Say the Least. A Young Pre-Engineer Drilling Wells in Timbuktu?

Yes, Timbuktu, in Mali. And we were not even in Timbuktu, but in a wide zone towards the northern portion of the settlement, the city.

I was, of course, in the Sahara Desert. A desert that saturates your senses, that fills your vision and even emotions. I was there, enclosed by sand dunes and taking care of a drilling rig. It turned out I liked drilling oil wells. It was amazing, unique. I felt that that is what I wanted to do.

"With Nothing Around Us"

I think we were 200 km away of any kind of settlement, completely detached from civilization of any kind.

From Paris to the Sahara!

Yes. We were right in the middle of the desert, with nothing around us. I started drilling, and the sensation was as if I was really in the line of fire as a young drilling engineer. I felt the energy of being at the center of the business. We were at the center of the action of the company. Everybody wanted to know our progress with the drilling bit. Every foot we drilled counted!

After that unforgettable experience, my company assigned me to an innovative project dealing with solar energy. You have to consider we were tackling this back in 1980, more than 35 years ago. In 1980, not many oil and gas companies were dealing with non-conventional energy; much less about solar energy. But I was there, in this project, aiming to energize a hospital with a complex system of generators powered by photovoltaic cells.

That town had no electricity except the energy created by generators. So, I was in charge of installing this huge complex of installations. I mounted the panels, the cables, the connections,…in summary, I instrumented the hospital. I was glad to see how, with this, we changed the life of all people connected to the medical services. We were reducing the costs, and providing reliable, continuous electricity to work, to cure, to sustain the medical treatments.

Where Was This Hospital?

This hospital was in Kolokani, a town that at the time had about 30,000 inhabitants, in Mali's Koulikoro Region, towards the southwest of Mali. Kolokani is the capital of about 10 rural communities, and all of them utilized that hospital. It was a great step up for the people of that region of Mali. It was a very good, rewarding experience.

The Norway Years

After Kolokani, I went back to Paris, and by this time, I had already graduated as an engineer. Total (well that was Elf at the time, but now Total) offered me a position immediately. My very first official job was in Production Methods. I was working for the commissioning for the processing engineering operations of a new field. That role taught me a lot about the technical specifications concerns, and the considerations of Health, Safety and Environment (HSE). I worked at the office for a while, and acquired expertise with very knowledgeable personnel.

Very quickly, I was sent to Norway, as Production Methods Manager. This assignment allowed me to become specialized in gas metering. Spending half of my time on the FRIGG field in the North area, that was the main source of gas for the UK through a pipeline to Scotland.

So, you started with offshore.

Yes. I started with offshore. After that, I was assigned to a project in The Netherlands, as deputy Operations Manager. Always in Elf-Netherland. It was the Northern Offshore Gas Transport (NOGAT) project, in partnership with Shell, to provide and bring the gas back to onshore. Very interesting.

So, You Started with Offshore

Yes. I started with offshore. After that, I was assigned to a project in The Netherlands, as deputy Operations Manager. Always in Elf-Netherland. It was the project, in partnership with Shell, to provide and bring the gas back to onshore. Very interesting.

The Growth of Angola

My career evolved from the beginning with multiples changes, and I changed focus as I moved countries, in different roles, very rapidly. Assignments were shifting me to new places. After my assignment in The Netherlands, I went to Angola as Operations Manager, in charge of a field producing 60,000 barrels of oil per day, and when I left, the production was 200,000. This was done in only three years. It was an amazing amount of work that produced astonishing results. We were able to more than triple the production!

There were 600 employees in my group, starting a platform every 6 months. We were coordinating a very complex operation, installing platforms, initiating construction, and dealing with drilling and production operations, all at the same time.. I am very proud of what we accomplished in Angola.

I enjoyed my assignment in Angola. It was a moment of tremendous professional growth, and the biggest achievement of my career at the time. I liked the sense of accomplishment, the complexity of the operations, and the huge experience I gained on a daily basis, with a big group of people.

"I Decided for Business Management"

After that, I had to make a choice: whether to continue as a technical operations manager or to switch to management in business. I decided to go for business management, although I experienced a lot of pressure by the directors to stay in the operational arena.

Was It a Difficult Decision for You?

Not at all. I was certain I wanted to have a voice in my company's business. I was convinced that by knowing the operational side inside out, I could make a difference in the decisions needed for the business, from the planning to the operational stages.

So, I was sent to the UK, were I headed a Joint Venture (JV) Management Department. The role was JV Manager, and I was coordinating 12 Joint Ventures, located all over the United Kingdom. I was managing the authorities as well as each partnership with different companies. For four JV's. It was not easy, let me tell you. It taught me a lot about different perspectives in business development and cost reduction.

After this assignment, I went back to Paris.

After Several Years. Finally Back Home!

Yes, back home. You know? I have spent in all my career working very few years in France. In total, out of 36 years in the company, I have spent less than 7 years in my country. I feel I have always worked as an expat. Adapting, fitting in, and understanding different ways of living and working. Different styles. You adapt. I have learned it by doing. You adapt!

During this period in France, I was appointed Vice-President of the Elf Affiliates in CIS (Russia, Azerbaijan, Kazakhstan, and Turkmenistan), Eastern Europe countries and China. It was one of my most remarkable experiences, from the point of view of multicultural experience.

You Have a Long Trail of Success, and in Several Roles. Lionel, What Learning in All These Years Marked You the Most and Why?

Lionel sits back and thinks about the question. Not an easy one, for a globetrotter like him, for sure. We notice he goes back and shuffles mentally through his past experiences. Finally he declares what was to become the core of the interview.

I think the most important thing was something that I realized when I was a senior Manager in my UK assignment. That element, that piece of realization, was that the people are the most important thing. For constructions, installations and technical projects, you always find a solution. There is always a solution. But the real issue, and what makes the success of everything, is the

PEOPLE, because from this side of the business, you need to be supported by people. I have found this to be true so many times!

Any Anecdotes, Lionel About the People in Your Teams?

There are uncountable anecdotes of why you need to motivate, concentrate and lead the people, your people in your teams, to steer the success of your projects and your company. There are too many to enumerate. I will share with you just the lighter side of this close connection that I have established with the people in my teams. I will share with you the close connection one establishes when truly caring.

Two Families Not Too Far from Each Other

This is one of those anecdotes…. When I was in Angola, there were employees in rotation 4×4, four weeks working and four weeks off. One of these guys, an extremely good and experienced maintenance guy, would use these rotation, commuter scheme to sustain two families. He had two families that he fully sustained, not too far from each other! To each of them, he was saying he was working in 6 × 2 and coming back home for only 2 weeks. He was effectively working 4 weeks in Angola, 2 weeks with one family and 2 weeks with the other one. People who have difficult personal situations, people you meet and with whom share personal challenges. I have met thousands of people, of all levels and cultures. You realize you have to lead all of them, and that the most difficult things to manage are your own emotions and your frustrations. If you do not manage these, you will not succeed.

Money Is Only Money

Manage your emotions. The rest is a piece of cake. There is always a solution. If you do not have a good people in management, you are dead. To realize that was very, very helpful to me. It was probably my most important learning in leadership.

I had the opportunity to meet tough people. Especially in CIS. To listen to Mikhail Borisovich Khodorkovsky, the CEO of Yukos at the time, was an unforgettable experience for me. He had a special insight for business and a vision for his country. He is now resident in Switzerland, after spending

10 years in prison in Siberia. All this political turmoil that ended with his imprisonment happened a long time after I met and liaised with him.

He was the CEO of Yukos, controlling a series of giant oil fields in Siberia, all under the Yukos. He was the wealthiest man in Russia and maybe in the world at the time, extremely rich! An amazing person. One of those fascinating guys you meet once or twice in your lifetime. He made a lot of mistakes. Sure! But he was fascinating.

I have been very attentive to mistakes. I always listen to what people say, and that helps me to realize if I am on a good course or if I should make adjustments. He was the CEO of Yukos, controlling a series of giant oil fields in Siberia, all under the Yukos. He was the wealthiest man in Russia, extremely rich! An amazing person. One of those fascinating guys. He made a lot of mistakes; but he was fascinating.

The Dolphin Project and Khaldoon

Ending my assignment in the CIS countries, I was assigned to the Dolphin Project, having the unique chance to work alongside Khaldoon Khalifa Al Mubarak. He was one of the most important entrepreneurs in the world and certainly in the UAE. He was only 26 when I met him the first time. I may say we built Dolphin together. Of course, I must acknowledge the immense support of Sheikh Mohammed and Christophe de Margerie, who envisioned this project.

Khaldoon is the CEO of Mubadala Group and also the Chairman of Abu Dhabi Executive Affairs Authority. He is also the chairman of the Manchester City Football Club, and this is only to give you a glimpse of the stature of this great man, whom I consider my friend. I learned a lot from him. He left a powerful imprint in my career. He impressed me when he was only in his early 20s, imagine that! The Gulf has indeed potent, powerful business men and women who have given me a glimpse of what is to lead from zero to the top of the class, when you adequately use resources and leadership insight.

Of All Your Achievements and Challenges, Which Were the Most Important for You and Why?

My most important achievement? Dolphin. Definitively! Because it was a fantastic project that has propelled and ensured tens of billions of dollars for Qatar, for UAE, for TOTAL. A pureblood win-win project. This project gave me the opportunity to get to know on a personal basis, beyond Khaldoon, a lot of very interesting persons in the Region. Dolphin has been my most important achievement so far.

About challenges, I must say I have never been demotivated by challenges. It is not my money involved, nor my life. If you want to do a good business, and good work, you have to remain very cool. Somehow, even detached. Remember I mentioned this at the beginning of our conversation? Control your emotions!

The biggest of my challenges was Kuwait. It was a convergence of a series of factors that make me catalog it as my biggest challenge. Of all situations I have handled at work, the combination of context and timelines made it extremely difficult to attain a win for Total in Kuwait, and hence I interpreted that as a big challenge, as I was the General Manager of Total Kuwait for five years. Nevertheless, I must say that in Kuwait I met people who impressed me. Nizar Al-Adsani is bright, smart, and a gentleman. Hosnia Hashim, because when you look at the conditions of women in the Middle East, she is one of the persons I respect the most. She had to be 10 times as good as or better than men in the same roles she has been. These individuals are climbing invisible, steep, difficult ladders to reach to the top. They are doing something for their country.

Some People Had a Special Influence in Our Careers. Who Was This Person for You and Why? I Am Referring to Your Role Models

I think I have an ability to spot what is good in each person. I will always remember this General Manager in the UK, he must be enjoying a peaceful retirement now, he was a one of a kind. What I will always remember is what he told me: *when you have people coming to you, always treat them well. Even if you agree or not agree, whatever you think of them. Treat well especially those you do not appreciate.* We were in a cabin, in one of the platforms, and this observation was provided as a comment to me after a difficult conversation with an operator. This was a key learning for me. I still use it consciously. Especially with the people I do not appreciate!

We laugh, and realize that is applicable everywhere, with everybody. What a lesson! And in an offshore platform? Leadership lessons are everywhere in the oil industry. No doubt.

Respecting People

What Are Your Personal Strengths that Propelled Your Success?

Lionel thinks for a while. He takes every one of our questions seriously. We have already immersed him in a personal journey of self-reflection, and I suspect he is enjoying it. He comes back with this direct assessment:

I think my personal strength is that I always respect people. I can see in ways that other people do not. I acknowledge this is not easy for all. The behavior has to be number one. You have to be extremely respectful. I have been negotiating with almost all kinds of personalities in the World: African, Russian, Americans, Middle Easterners, Asians, but from my experience, you have to respect them, even if they treat you like trash, and I have been treated like trash many, many, many times.

The Oil Is There

What Are the Challenges the Energy Industry Will Face in the Future?

In the future, in the past, and today, it is always the same story: the oil is there.

It is not the lack of resources. There is not a lack of hydrocarbons, but there will be a lack of good human resources. Human resources in the oil sector will be more and more difficult to find because there will be more and more regulations to follow, and more complex technologies to be mastered.

Adapting to different cultural environments, which is not always easy, is another challenge. Now that the industry is incrementally global, and talent is shifted across country and continent borders, it's important to adapt to help where they are needed the most.

At the end of the day, you have to be able to respect people with whom you realize you may never be able to agree on anything.

Do Not Let Anyone Move the Joystick for You!

What Message Would You like to Send to the Young Generation?

First, enjoy your job because, after all, it is where you dedicate the vast majority of your time. The most important thing that you do not see very often is to know what you want to do. Create your vision! If you do not have a vision, it will be difficult to succeed. Never let the organization to do the choice for you.

I like to say that you do not let anyone move the joystick for you!!

The worst thing that I have seen many times, is that the vision the people have for their careers, is too often much higher and ambitious and greater than the one their bosses have envisioned for them. Most of the time, it is very different from the reality. I deal with this fact every day.

If you know your weaknesses, you will know what to do to overcome them, and match your vision.

A Shared Selfie

- **Your favorite word:** It is not a word. It is a phrase. "Never give up!"
- **A city:** Paris and Rome.
- **An important person of your preference:** My family.
- **A favorite landscape:** The one in my house, in Amboise, France. But also Madagascar, Cape Town, Ashgabat, and Turkmenistan… so many beautiful landscapes.
- **A historical figure:** Mark Twain, I think the guy was really amazing!
- **A personal happy moment:** I have had so many! Thousands of these moments. I am very optimistic. I remember one funny moment in the UAE, a meeting with 15 men, all named Mohammed. It was a surrealistic moment, flavored with laughs. To be with Caroline, my wife. My children. And to be with my older daughter.
- **Your favorite food:** Spaghetti alla Bolognese. And also there is a restaurant in Paris, where they serve veal liver, a delicacy that I crave. At least once in a lifetime, you have to try it.
- **A song:** Sympathy for the Devil (Rolling Stones). The most amazing song ever composed.
- **A singer:** Bruce Springsteen. He is the King of R&R, the Boss.

- **Who supported you the most?** My wife, Caroline, is truly an exceptional person. My mom has a special place for a certain stage in my life. And I must mention him, because he is a very good person, Stephane Michel.

Post-scriptum

Towards the end of the interview, an unexpected facet of Lionel was uncovered. I was mentally summarizing if I had covered all points, and I directly asked him if there was anything left out about his work story, that we did not capture.

I was surprised to see him vacillating, just for a few seconds. As if he wanted, and did not want to let me know about something. I insisted. I intuited there was something more.

And then, Lionel provided this revelation as a gift: Lionel is an avid collector, expert and unconditional fan of Star Wars memorabilia, movies, and figurines. Who would have guessed that? It was refreshing and even reassuring to realize once more that leaders are multifaceted, and many have multidimensional interests. Lionel enjoyed focusing his never-ending energy on Star Wars. Wow!

Star Wars! I like It a Lot Myself. It Was THE Movie to Go in the 70s

Yes. Star Wars, I am an expert of STAR WARS, with a huge collection of figures, posters and toys. This is very important in my life. I started before entering in Engineering School, at a time when I was a part-time journalist, as a critic for fantasy films, for the *Fantastic Screen* magazine in Paris. I specialized in Horror and Sci-Fi movies. At the time, the Italians were very famous in that sector. So, as a sci-fi critic, I had the chance to be at the premier of Star Wars, and I was blown away.

Have You Attended the Comic Con in USA or in Dubai?

Not yet. But I will! When I retire, I will go with my son. That is for sure! That is something I want to do before I die. I love all about Star Wars. I love the soundtrack of Star Wars, you cannot compete with John Williams. Even with George Lucas: the Script was unique. The story is so simple and so amazing at the same time. The characters!

I was in shock to learn that this man had a true passion for Star Wars. It was amazing. I never, truly never expected this. We share anecdotes of this past summer, when Star Wars 7 was released, and I tell him how I was one of the loudest who cheered Princess Leyla (old princess Leyla), on her (for me) unexpected appearance in the Episode 7, at the Houston Theater where I saw it. We were two children talking about Star Wars. As if we were again in the 70s!

It is a movie from the 70s and the face of my son when he saw it for the first time, was like my own face.

You have three main movies: Episodes 4, 5, 6, then three movies not of the same quality. After The film 7 is a very good continuation.

Hinda Gharbi

"I was always clear on what I wanted to do".

A Glimpse

A Tunisian female engineer is building an undeniable trail of success in the oil industry, with her many achievements in the most important oil and gas fields of the world, in a leading role for one of the top-ranked oilfield services companies. That engineer is Hinda Gharbi, a tiny woman who has built for herself a high reputation. Hinda has broken paradigms about women at the field in the oil industry's offshore operations, raising eyebrows of colleagues and strangers, from her early start, having to jump into helicopters to reach her worksite at the drilling platforms at sea, to lead well logging operations in what was many times a first, for those operational crews.

Hinda is now the President of the Wireline Division of Schlumberger's portfolio of service and product offerings with a global responsibility, based in Paris, France. She was President of Schlumberger's Asia region in 2010, based in Kuala Lumpur, and before that, Vice President for Health, Safety, Environment, Global Citizenship and Global Regulatory Compliance, based in Paris. In earlier assignments, Hinda was Managing Director of the Central East Asia market, based in Bangkok, and responsible for the operations of Schlumberger in Thailand, Vietnam, Myanmar and Bangladesh.

- Master's Degree in Signal Processing and Engineering Degree in Electrical Engineering from the Institut National Polytechnique de Grenoble (INPG) in France, 1996.
- Joined Schlumberger in 1996 as a wireline field engineer.
- President Asia Region, 2010.
- President Wireline, 2013.

A Personal Snapshot

Hinda has that special touch of a leader who cares. Care for others and for excellent technical quality is evident in all she does. Hosnia started to liaise with her from the time she was President of the Asia region, and during these last years, the contact enabled Hosnia and I to incrementally know her better, as a truly progressive and insightful leader.

The environment of the Middle East is very much networked with relevant figures in the oil industry all over the world. Undoubtedly, Hinda was already a relevant figure already back in 2010, when we met her. Hinda does not make you feel the importance of her role; and on the contrary, she is extremely accessible and approachable, inviting an open exchange. It is remarkable how at the distance she can still find time to enable the establishment of sincere and caring networking efforts. We are privileged to have Hinda among our enduring liaisons.

Arranging the Interview

Hinda Gharbi had come to Kuwait in several occasions, for business reasons, always leaving a great impression in Hosnia. But it was her presentation at our Professional Women Network (PWN) Summit, in October 2016, that triggered our wish to include her in this compilation about resilience. Her participation in the PWN's Summit, for the Kuwait Petroleum Corporation and its Subsidiaries, had enlighten the audience about her particular story of success, one of many pioneering "firsts" in the oil industry. We wanted to learn more about her roadmap, about her own stairs to leadership. We were curious about how she climbed every step upwards and onwards.

An accomplished leader, she replied to our invite for a conversation about leadership and resilience with a splendid and swift "yes!". Then, we tried to reconcile her busy schedule with ours, but could not find a common time and place to meet in the following four months, so we agreed the conversation would take place over the phone. It worked out beautifully.

The Progressive Girl Who Became President of a Global Business

"I Was a Progressive Girl"

We know you are a role model for many in your company, and in the oil industry in general. Now, you are more known in Kuwait, especially by the young professionals who attended your presentation at the Professional Women Summit, in October 2016. We want to know about your beginnings and how you progressed your career.

Thanks. I must say I admire very much Hosnia, not only for her own merits and achievements, which are many, but also for her efforts in advancing women in Kuwait.

Hinda wanted to know if she would only speak about her career, but we explained that we also wanted to know how she became to be who she is today, from her childhood. This is how we discovered many details that were hidden, but that account for her resilience and story of success.

I was born in Tunisia; I am Tunisian, and that is why speak Arabic. There, I studied my Baccalaureate, and I guess I should tell you I was always good at school. I was raised in a way that shaped me as a progressive, tenacious and resilient girl. My mother ensured we could look after ourselves. She taught us to be highly independent.

I was sent to summer camps in our own countryside, in Tunisia, but also to Morocco from the time when I was seven years old. I am convinced that being far away from home at such a young age forges your character. Additionally, I also participated in a lot of sports. So, all these elements in my upbringing made of me open to change. I was resilient from that moment, most likely. Then, my tenure in France for my studies made me even more resilient.

In the Top Ten Percent of the Class

Tell us how the idea of studying in France came to be.
I always knew I wanted to go overseas.

In Tunisia, the government would sponsor studies in Europe for those top students of the high school system, and would send them to Germany or France to get an engineering or science degree. To me, pursuing my higher education overseas was a path that would happen only if I did well at my grades. So, I had to keep good pace and excel at exams and grades to be selected, as only the top ten percent or so were chosen, a very select group from around the country, of which I wanted to be part.

It was not easy to get to be in this group. But with the strong support of my mother, I applied myself to my best efforts. She encouraged me to apply for a scholarship from the Tunisian Government to study overseas.

"Eulim"

My mom passed away this past October. She was a very forward looking lady! Learning was very important for her, and it comes from her strong faith, where learning is a natural path. Learning in Arabic is called "eulim" (science-knowledge). She wanted me to learn, always encouraging me to better myself. My mother was not highly educated, but she was a creative and curious person. So for her, when I got my scholarship it was not a reason to worry, but to rejoice. It was my mother who worked to sustain all of us, 5 children.

I imagine other families surrounding yours and perhaps your own distant-family members would worry about you going away alone to study. During those years, this was not common practice, right?
Absolutely! Everybody around my mother was concerned, but she did not share their concerns. She used to repeat "I know how I raised my daughter, and I have no worries." She was definitively not traditional.

I chose engineering. In France, classic preparatory for engineering is part of a special system, one of intensive preparation for engineering schools, including lots of math, physics and science and chemistry. We were among the top people from all of France, and it was the first time I went to study far from home.

Not easy at all. What kind of strength did this separation from your family and this heavy study regime forge in you?

I would say my whole experience in France made me stronger, forged the tenacity and resilience that helped me subsequently work in the field.

Studying in France

How was your experience as a student in France? Were you with a host family?

No, I was in a dorm. We were only a few girls from Tunisia, and it was hard, because on the weekends, all our classmates went home, but we stayed. We stayed together, and studied. We had to! It was super intense, with exams every week.

I graduated in electrical engineering, and I also attained a Master Degree in Signal processing, both in Grenoble. My alma mater is the Institut National Polytechnique de Grenoble (INPG), for short. It was near the Alps, a very beautiful location.

What kind of work path were you envisioning at the time for yourself?

Numerous electrical engineers would work in electrical engineering companies. Some with my Signal Processing qualifications would be hired by seismic companies. I had the opportunity to do an internship in my second year of engineering school, and I then discovered the incredible Oil and Gas industry.

Hinda, Schlumberger is very important in your life, and the platform for your career. How did all this liaison start?

Schlumberger was a very visible company on French campuses. They had an impressive presence in recruiting events at the top engineering schools across France. Their message of technology and people development resonated with me. We had to do an industrial internship in the second year of engineering, and I chose one with Schlumberger as a field trainee in the northern region of the Netherlands.

The Eye-Opening Experience in the Netherlands

My experience with Schlumberger in the Netherlands was an eye-opener. It was an absolute discovery journey for me. The field engineer job was amazing. I closely observed those engineers I was shadowing, and admired the variety of tasks they accomplished in a single day of work. I was amazed how quickly these young engineers were put in charge of multimillion-dollar equipment and a crew. My mind and heart were captured by the multitude of things they were accomplishing, the advanced technology they are using, the level of responsibility and accountability entrusted to them so early! The technology, especially for me, was mind blowing. I was amazed.

After this internship, I wanted to be part of this very unusual international career path that Schlumberger offered. I thought it was right for me. I became enthusiastic about the idea of working in a technology-driven industry, and moving to different countries.

I graduated in September 1995, when I presented my thesis for my masters. During my last years in university I was able to study for a master at the same time as finalizing my engineering degree. I wanted to increase my chances on the labor market so I attempted those two degrees at the same time, resulting in a heavier studying load.

"If You Could, You Should"

And then, I got an offer from Schlumberger to join as a field engineer. I remember very vividly that I signed the contract in Paris, and it was February 5th, 1996.

How did your family take your decision to work in the oil industry?

They supported me. Especially my mother, as for her it was a great opportunity, and she would say that if you could do it, you should. So, I could and I did it!

They first sent me back to Tunisia for about a month, in what they called pre-school training, to learn the basics of field work in oilfield services. After my month of training, I was sent to Edinburgh, Scotland for a 3-month training. It is a school for all new hires of Schlumberger from around the world. The aim was to train to become a field engineer. After that, I wanted to work in another place of the world, not in my native country; I wanted to discover new places. "Do you want to discover other places? No problem!" my supervisor told me. "We have openings in Nigeria". And I was sent to Africa, to work in offshore operations.

Nigeria, the First Expatriate Assignment

I remember very vividly the day I arrived for the first time in Port Harcourt. There were numerous armed guards at the airport, and I recall the humidity and heat, and the atmosphere of a tropical country. All was new and different for me. I stayed in Nigeria three years, working exclusively offshore.

What did you like most of this initial phase of your work?

I enjoyed it, because of the nature of the work, which included some of the most prolific oil fields in the world where advanced technologies were used routinely with the International Oil Companies (IOCs) operating offshore. It was a massive exposure to complex jobs on the rigs, and sometimes involved staying at the offshore facilities for weeks. I had my own crew, and at the beginning was trained by senior engineers. It was a very intense schedule, and there was lots to learn! It was very interesting and rewarding!

At that time, I'd say that my English was hesitant. That was an additional immense challenge, as everybody else on the offshore operations were from different countries, and communication forcefully had to be in English, so I needed to improve my English fast. But many of those working with me were really supportive, boosting my confidence and allowing me to progress.

"Does Your Father Know?"

During a month of training in Tunisia, some of the workers in the remote fields where I was part of a crew performing operations, were surprised by my presence there and would ask "does your father know you are doing **this** work?" emphasizing "this". They were implying, of course, that I was alone in the rig, in hard duty assignments, and committing to night shifts with the crew. They would insist: "Does your family know?" or the more common "do you have brothers? Do they know you are doing this kind of work?" I would answer that yes that my family knew. And yes, they knew, and they were proud.

I know those gentlemen were asking because they were surprised. I was the only female engineer on those fields, and living arrangements could be tricky.

During my assignment in Nigeria, I was participating and given responsibilities, including dealing with customers. I liaised with the Majors (Mobil, Elf, ENI…) that had operations in offshore Nigeria. You had to deal with your crew to manage and execute your technical work. In those years I met my husband. He was an Australian field engineer, and has been my husband for seventeen years.

The Call to Leadership

Hinda, when did you realize you were a leader?

That is an interesting question indeed. I don't know if I knew I was a leader, but I would say that I was always clear on what I wanted to do. I was determined; I wanted to accomplish things, to better myself, to discover new challenges. There was no doubt in my mind that I was driven to be the best, to be able to make my mother proud. She worked very hard to help all her children. I wanted to progress, and to deliver results.

I was focused and on top of what needed to be done. I was working to achieve my goals. I was geared towards materializing what I wanted to do. I was fortunate to be clear about what I wanted to do. I quickly realized Schlumberger was the kind of company in which I wanted to be.

It Is a Tough Work

What did you put into in your leadership backpack from Nigeria?

I think that something important I learned in Nigeria was the sense of community and the great camaraderie in the oil and gas. It was a special bond, which you can feel. You are with all these people that you have never met before, but you share with them tight schedules in a very high risk environment. It is an amazing bond.

You really get to see how people are. That was an incredible experience, which made me grow as a person. You see the good and bad side of things; it is not always straight forward. This was tough work, far away from family that marks you.

Three years in Nigeria is a very long time in pretty difficult operational schedules. Towards the end, security concerns increased, and I was about to finish my training program. During this tenure in Nigeria, I completed the structured learning training program that Schlumberger assigns to all new engineers, and was promoted General Field Engineer, and sent to Houston.

Three Jobs in Three Years

In Houston, I executed three different jobs in the time I spent there, exactly three years.

My first assignment was as Field Test Coordinator, in charge of testing and promoting a specific technology. I was basically the link between engineering and the field, to test and introduce technology. This was an extraordinary opportunity to get to know the world of engineering and technology development. I needed to translate the technical features of the developed technology into benefits and value that our field operations could use in our clients fields and projects. In this sense, I gained an edge in communicating with very different teams. Those technology-driven with those operationally-focused. I did that for almost a year.

Innovating Recruitment

Then, I was asked to work on a novel project to build an internet recruiting platform to recruit engineers in the US. I was based in Houston, and additionally I had the opportunity to coordinate the recruitment in some top universities in Texas, Michigan, Pennsylvania and many others around the US.

I worked in a team, with another person who was the senior project manager for this pioneering idea. I was selected to do this project. The system we developed was in place for over 10 years, being renewed just recently.

My work on recruitment was a completely different exposure opportunity. I was in charge of hiring new engineers for the company. It was very interesting, challenging, and a very high responsibility. This truly opened my eyes in relation to understand how the DNA of Schlumberger was built, and how to communicate it properly to the young, fresh graduates to be appealing to them. All of this occurred during the first Internet bubble.

Oil and Gas was not appealing to fresh graduates at the time. We were competing with all the new Internet companies, and our industry was struggling with very low oil prices, giving an impression the industry was in its twilight. We had to explain to the students our values, what we stand for and why it is an industry vital for our economies. We noted that it is extremely technology driven, with sophisticated project management and a global footprint.

The Gulf of Mexico

The third job of my Houston tenure was to manage one of our operating bases in the Gulf of Mexico, an important region for our business.

How did this happen? Tell me about the shift back to operations.

I was asked to take on this new role to manage operations out of Webster, Texas. I have been always found operational work to be appealing, and I was thrilled to be given this opportunity. I had to cover the West of Houston, in particular, offshore logging, which is a high volume of offshore operations. It was a multimillion dollar operation, with over 50 employees. I was responsible for the financial results, health, safety and environment (HSE), operations execution, training of the employees and communication with the clients.

That was your first management job. What did strike you about management?

In my previous roles, I did not have people to manage. This first management role was very humbling. I was happy to be entrusted to manage such an operation at an early time of my career. I have been then only five years in the company. I led all these experienced and knowledgeable engineers. I was supporting all the aspects of running such a business unit, supervising two other field managers who coordinated and directed the execution of operations. I was proud to work with such people and to earn their respect and trust.

Here, in this job, I realized what it meant to be in charge!

Communicating the Way Forward

All these people's lives depended on me. It was quite an experience. It really taught me to communicate. I learned listening, especially. I learned to understand the direct and indirect messages from my people. I learned to build the right teams to do the work properly, and to trust my teams.

What weight would you give to the communication in this operational role, your first managerial experience?

People would get lost if they do not know where they are going. So, for me communication in leadership has a primordial importance, in providing a strategic direction.

I think leadership is to be able to let people understand where you are going as a team. It involves setting the path, and to do that constantly. You share with your team members what is needed. If you cannot communicate, it is very hard to lead.

The New Shift

After my third year in Houston, I received a call from my boss, who was an amazing manager and mentor, an American gentleman: "You are getting transferred to Aberdeen, in the UK. You will be the Human Resources Manager for the Geo-market there, for all the business".

It was a difficult decision for me to understand, causing me to reflect mistakenly whether or not I had a people management problem. I realized later that the company just wanted me to know how the inner wheels of the heart of the organization works, and to gain experience in understanding people management challenges, by handling the human resources department. It was a moment of courage and resilience. I was discovering a new world.

Together

And in all these movements, how did your husband and you maintain the family unity?

I have been blessed with a dual-career journey, where the company facilitated my transfers and those of my husband, who also works in Schlumberger, to be together as a family in the same location. And of course, it's an advantage to have an understanding, wonderful husband, who has envisioned how to advance his own career along mine. This is our own personal teamwork, which has worked beautifully for us and our two children.

For this particular move to UK, for example, my husband was offered an opportunity to complete a one-year master degree in an Edinburgh university, in real-time production optimization project, sponsored by Schlumberger. It was dual career policy in action. They knew they had to give him an opportunity. We have always moved, and thrived together!

The Young. The Small. The Lady. The Expat

This move to the UK was in summer of 2002. I went to Aberdeen, and became the HR manager for the UK and Ireland, responsible for the human resources function of about 1,800 employees. My team in HR was about 20 people, and I had to liaise with numerous managers, peers, who were operational, and responsible for all types of operations. These were very senior people, whereas I was the junior engineer, with only six years and four months in the company, doing a very critical job for the organization. I had, at the time, an excellent female manager, who was a great support during my first months in the job.

It was clear I was not an HR professional, and I needed to gain the trust of my colleagues very quickly. I put the effort to understand the fundamentals of

HR, and worked with the managers, spending time with them, and finding solutions for the key things with which I needed help.

What did you learn from this experience?

Working in HR gave me an incredible view of how a company actually works and the important place people have. Most importantly, I learned how to spot talent. In any organization, people are the key. This understanding is vital for companies to excel in what they do and to reach new heights. I understood this element very early in my career, and this kind of insight has sustained my leadership growth all along my career since then.

The genius of Schlumberger is the ability to give employees the opportunity to do different things along their careers. I benefited from this approach, and I consider that this assignment was an opportunity to grow, to have great clarity on how the company operates internally.

The Downsizing Experience

It was the first time I had to handle downsizing decisions.

That is hard. How did you handle that?

It is always very hard to let people go. One understands the economic reasons for having to downsize, but the important part, however, is to ensure that it is always done with professionalism and empathy. It is a complex and intense process, and I needed to learn how to handle the emotions and pressures that come during these times.

It was important to remain resilient, and the help from the HR team and from my managers and colleagues helped greatly.

At the end of this assignment, I had my daughter. She was born in Aberdeen. I took six months of maternity leave.

A Global Responsibility

Upon my return from maternity leave, I was asked to move to France, in wireline, in the role of New Technology Manager. I was responsible for the business side of the chain of technological developments for wireline, from the technology definition, to the introduction to the market. It was a global responsibility. It was my first global role.

I absolutely loved it! I particularly liked how we worked to find value for our customers, and how we translated the value of technology for the business. Working with our engineering people and technical experts in our customer's organizations was very rewarding. I was working with research for the first time, and I was again extremely impressed by how these scientists think and innovate.

The biggest accomplishment was that we introduced several technologies very successfully to the market. This is another segment of my career when I felt I became a translator, from engineering to customers, translating what our technology brings into value for our customers projects. You see? Communication is so important. I am still amazed to see how the power of communication shapes the attainment of the goals you set for a group, and for a company. For the collective way ahead in an organization.

After this extraordinary experience in the technology development and marketing area, I had my second child and took another six months off for maternity leave.

Upon my return, I was promoted to general manager, based in Thailand, and responsible for the business in Asia Myanmar, Thailand, Vietnam and Cambodia, and all business in that region.

I stayed in Thailand for two years and a half. In 2008, the oil crisis hit us hard, and I helped merge my business unit with another one in Asia, making my own job redundant.

Did You Eliminated Your Own Job?!

Yes, I did! But to tell you the truth, I was told not to worry about being left out. On the contrary, at the end, I was asked to move as Global Vice-President for HSE and regulatory compliance, working with Schlumberger's chief of operations. This was an extraordinary promotion to a role from which I learned a lot.

The HSE assignment was truly interesting, as you have to support businesses around the world as you focus on providing them with the right tools to help them manage safety, security, health, and the environment. In a large organization like Schlumberger, there are emergencies and events on a regular basis. It was very interesting and formative, and I had to think what value I would bring to the function, as I was not an expert in HSE. I applied myself in every move to become knowledgeable in the subject matter at stake, every single time and find where I can add value.

And you shifted a lot among functions, geographical regions, roles,…

Yes. I definitively had to adapt rapidly. I think that is part, an essential part of my leadership.

After only one year in my role of VP for HSE, I was asked to go back to Asia, to re-create the Asia region, and re-shape it in organizational terms. So I moved to Kuala Lumpur as Asia President for all Schlumberger businesses in Asia, a completely new office. I put together the team that would oversee our businesses from China to Australia. I enjoyed that role a lot. The customers, the activity growth, the amazing talent of people in the region, so many things to do.

I really enjoyed living in Asia, returning there after three years of absence from the region. Then, in 2013, I was moved to Paris, for a global role as wireline President, which is my current responsibility.

The Next Fifty Years

What is the most difficult aspect of being a wireline president?
This business has been around for the last 90 years, and is a core business for Schlumberger; the founding brothers of Schlumberger were engaged in the first electrical well logging! So, for me the challenge ahead is to continue in a growth path, to continue to lead in wireline operations. To be here for another fifty years.

Are you expecting another call from the CEO to yet another higher-ranked role?
We will see!

The laugh of Hinda fills the moment, and I happily join her, as we have dig into the difficult and joyful moments of her career in this extraordinary summary of success. I was impressed by Hinda's adaptability, her willingness to move to unknown expertise areas, and her fearless attitude! All these qualities are solid grounding to build a resilient leader. She faced many challenges, and not only survived in attaining and surpassing her goals, but in transforming herself with endurance and commitment into the leader and role model she is today.

We expect to see more from Hinda in the coming years. Whichever path she will take, it will surely be an interesting one, motivating for the young generations of professionals in the oil and gas industry.

Difficult Replication

What are the challenges the energy industry will face in the future?
Undoubtedly, one of the major challenges will be how and whether unconventional resources will be exploited around the world.

The success in the US of non-conventional oil from shales is going to be very difficult to replicate outside of the US. I do not know if we can even replicate it.

A Shared Selfie

- **Your favorite word:** Possible.
- **A city:** Paris.
- **An important historical figure:** I love the story of how Queen Dido founded Carthage, a historical iconic city in what is today Tunisia.
- **Your favorite food:** My mother's couscous.
- **Your favorite color:** I do not have a favorite color. I like them all. If forced to choose, maybe I would pick green.
- **Who supported you the most in your life:** My mother.

David T. Donohue, PhD, JD

"My guiding principles are innovation, excellence, and hard work".

A Glimpse

Every two minutes, a person opens an IHRDC e-Learning course somewhere in the world and learns what and where they want to learn. IHRDC is one of the successful entrepreneurial achievements of Dr. David T. Donohue. More than 15,000 people in the oil and gas industry have attended some of his company courses, to expand their knowledge or skills. He started IHRDC in 1969, when he was in Law School. Today, it has a broad array of innovative training products and services with global customer reach.

Dr. Donohue also achieved an entrepreneurial breakthrough in the gas business, when he developed the first independent underground gas storage field in the United States. He envisioned a need to assist US gas distribution companies to meet their peak gas demands. He applied his technical, legal, interpersonal, creative and risk-taking talents to achieve his goal of becoming an owner of the facility.

Dave is an active volunteer and elected member of local government boards. He supports charitable causes including endowing faculty chairs at both Penn State and Boston College Law School, and is active in professional societies. His open personality has allowed him to reach many professionals, touching them and leaving indelible imprints. He is a role model of the best fiber there is.

- Basic Engineering Studies at McGill University in Montreal.
- BS (Honors) in Petroleum Engineering from the University of Oklahoma.
- Ph.D. in Petroleum and Natural Gas Engineering from Pennsylvania State University.
- J.D. degree from Boston College Law School.
- He began his career working for Imperial Oil, ExxonMobil and Gulf Oil in Canada and the United States in operations, reservoir engineering and research before joining the faculty at the Pennsylvania State University.
- 1969, he founded IHRDC while in his first year of Law School. In 1974, he founded the Arlington Group, developing the first independent underground gas storage field in the United States.
- Joined the SPE in 1959. Recipient of the Cedric Ferguson Award, Distinguished Lecturer during 1990–91, Distinguished Member in 1987 and Charles F Rand Memorial Gold Medal in 2016.
- The Pennsylvania State University recognized him as an Alumni Fellow in 1999 and the College of Earth and Mineral Sciences awarded him with the Distinguished Achievement Award in 2003. In 2009, he received the William Kenealy SJ Alumnus of the Year Award, Boston College Law School.

A Personal Snapshot

Dr. Donohue is admired for his enthusiasm, entrepreneurial spirit and success in the training field. He exemplifies resilience as he continues to work full days at the age of 80. He is related to the career of Maria Angela in several ways. An opportunity occurred in Caracas in 2001, when Dr. Martin Essenfeld, Dr. Donohue's former student at Penn State, was serving on the Board of Directors of PetroUCV, a partnership between the Universidad Central de Venezuela (UCV), and Petroleos de Venezuela (PDVSA), and introduced Maria to Dr. Donohue. Dave was subsequently invited to partner with the goal to craft a Master of Science degree at UCV partially based on

E-Learning. The proposal was accepted but the implementation faded away in the turmoil that disrupted the Venezuelan oil industry. Maria Angela had other interactions with Dr. Donohue when she was active with Halliburton, Kuwait Oil Company (KOC), and SPE, and they developed a strong liaison, launching training programs and competency assessments.

Arranging the Interview

Maria Angela asked Dr. Donohue if she could interview him so as to capture his learnings on leadership. He was very willing and arranged a Face Time session with her in Kuwait and him in Boston. A very modern method, for one who was obviously not hesitant to use technology for this aspect in his career.

One of Those Things

Tell us about your career.
My mother was from Pennsylvania, of German background. My father was an Irish dentist from Montreal. I grew up with a strong father and mother and four siblings in the Montreal suburbs. My father was very outgoing and made friends easily, and my mother was a natural leader, very organized and hard working. I have many of their combined characteristics.

How did you become a leader?
I would say there were three elements into this.

First, self-confidence: I always felt that I could lead and was often asked by my peers to assume the leadership role. I went to a small Catholic school, was captain of the High School Hockey Team, the Football Team and President of my High School Class. People would say, "Dave can do it", and I did. I also graduated at the top of my class.

Second, capacity to work hard: I really like to work. I began delivering morning papers from the age of nine. At the age of 12, worked with my grandfather, an oil producer in northeast Pennsylvania. When I was 16 years old, I worked in remote northern Quebec and Labrador loading airplanes and as a brakeman on an iron ore railroad. The money I earned gave me a sense of independence and allowed me to pay my way through college. It gave me a great deal of satisfaction and it was fun!

Third, an understanding of what I wanted in life: My parents separated when I was 12 years old. I understood then, that everything that I would achieve in life would depend on a successful education. I dedicated myself to learning as much as I could to open as many avenues as possible. I graduated first in my class in high school and was admitted to study Engineering at McGill University. My goal from an early age was to work in the exciting petroleum industry. Engineering was a natural; I liked science and mathematics. I paid my way to McGill University to study Engineering.

Give me more examples of how did you know what you wanted in life. This is not simple. Is it?

I took a number of paths to discover my "natural giftedness." An uncle said I should study Petroleum Engineering like he did, at the University of Oklahoma (OU). I studied basic engineering at McGill, and transferred to the OU, a tremendous change in culture for me! But I adapted, worked hard, made many new friends at OU and graduated with high honors. My first major goal had been achieved!

During the summer months of my junior year at OU, I worked for Imperial Oil in Edmonton, Alberta as a reservoir engineering intern. We used Friden calculators and accounting spreadsheets because computers were not yet available. I returned to OU for the fall semester, but soon ran short of money, so I asked Imperial Oil if I could return for nine months to a position in operations. They assigned me to be a Production Engineer. I realized that I preferred Reservoir rather than Production Engineering.

Why leave? There are people would have fought to work at ExxonMobil for a lifetime.

Nine months in operations were sufficient for me to know that I did not want to be in operations. If I wanted to be a Reservoir Engineer, I needed more education in reservoir engineering theory and mathematics. I considered a number of graduate programs, and accepted the offer from Pennsylvania State University that allowed me to study in a Reservoir Engineering program. In 1959 I graduated from OU, spent six months working in research for a drilling bit company in Houston and three wonderful months travelling throughout Europe, before beginning my studies in Pa with much excitement.

During my MS, I became interested in an evolving area of petroleum technology, *Reservoir Simulation*. I decided to continue to the PhD so that I could work in research in this exciting area. I knew that I should either stay at Penn State or move to Stanford. Penn State made a very good offer. My study led to a PhD thesis on multiphase flow in porous media with mass transfer. I finished my work in 1963 and headed to work in Imperial Oil's research laboratory to set up a Reservoir Simulation Group.

When I interviewed at Imperial Oil a year earlier, I told the Director of the Research Lab that I wanted to become actively involved in Reservoir Simulation and its applications. He said, *"You can come here and if you are good, you can lead the group. If not, you can still work in your area and someone else will lead."* The Director died three months before my arrival, and the new Director sent simulation work to Houston. I could have accepted other job opportunities that would have been more challenging.

Who Is This "Management"?!

I stayed on for a year, working on heavy oil simulation and analysis in both Calgary and in Tulsa. I worked with upcoming specialists, including Keith Coats, who became an industry giant in Reservoir Simulation, and Lee Raymond, who became ExxonMobil Chairman. I soon became disillusioned with working for a major corporation. It seemed like decisions were made by "management," some amorphous entity way up high in the organization. Once I had a suggestion to enhance heavy oil production, which was enthusiastically encouraged by my boss. Later we heard that "management said no." "Who is this management?" I asked. The bureaucracy was not very appealing for a hands-on person like me.

When Penn State invited me to join the faculty I quickly said "yes". Upon reflection I realize that I was not happy in a large corporate setting – it was too restricting. I needed to have more individual freedom.

Prima Donna Profiles

I joined the faculty of Penn State in 1964, and taught there for four years. It was wonderful opportunity with challenges to teach new programs, advise students, guide others and do your own research and publish.

Teaching is freedom personified. Dr Farouq Ali and I re-designed the graduate program for the Petroleum Engineering Department. This transformation meant that we then had to develop many new courses related to new developments in technology, especially reservoir simulation. I spent a summer at Gulf Research working on interphase mass transfer during the injection of LPGs into a light oil reservoir. My fellow researcher was Dr. Harvey Price. We won the SPE Franklin Award for that innovative work.

I was in Research and Development (R&D), preparing technical papers and presentation, but I felt all was a repetition. I knew I could not spend 30 years teaching.

My younger brother came to live with me prior to beginning his studies at McGill. We made a small investment in a rooming house. We purchased more apartment buildings and began building single family homes. I was the financial, budget and legal expert for our joint venture.

I really liked business, but could not continue doing what we were doing without more education. I combined the two things I really liked: (a) I would teach adults what they needed to know to keep up with developments in the industry, and (b) go to graduate school. I would move to Boston, which has a culture similar to Montreal, attend law school and start a company that became known as IHRDC!

Do Not Tell Him What to Do

I am happiest when I have the freedom to make decision on my own or with a small group to move ahead with a plan that is innovative, and has a high likelihood of being successful. Once the plan is set I am impatient to get things done. Sometimes I am wrong and freely admit it when it occurs.

I am not very happy in a situation where people tell me what to do.

How did you build your sense of confidence?

Confident is an inner behavioral trait reinforced by my mother and father, coaches, teachers and the successes in sports and jobs as I was growing up. When I was studying at McGill, I was told by elders that, if I worked very hard for 25 years for a company, I might become a senior manager. When I spent summers with my grandparents, I was told that I could become anything if I had a good plan, worked hard and had a strong education foundation. I was impatient to get started but had to find my way! The joint venture business, learning to develop curricula and teach at Penn State started me on my way.

30 People in the First Course

I entered Boston College Law School in 1968. My new wife, Pamela, and I lived happily, worked hard and began to build a family. I needed to generate revenue and so I spoke to Dr. Paul Root, about teaching a program for engineering graduates.

I formed the forerunner of IHRDC. We decided to teach a three-week program in Germany on how to write software programs from fundamental to advanced applications. We had time sharing computers, and we were able to arrange for Honeywell-Bull to deliver the lines to us in a hotel in Germany.

People from Japan, USA, Middle East, and Europe enrolled. It turned out to be an invigorating three weeks. IHRDC was off and running!

Boston: A Magnetic Appeal

I graduated from Law School in 1971 and passed the bar exam that summer. I liked Boston, and soon realized that developing businesses was more appealing than preparing contracts in a law firm. I would stay in Boston and expand IHRDC, but take on the additional challenge of building underground gas storage facilities.

Gas distribution companies serve residential and commercial customers who need large quantities of gas in the winter months to heat their homes and businesses. My plan was to find gas fields near the end of their useful lives in the closest basins, convert them to storage facilities so that we could serve these companies by allow them to fill the underground reservoir in the off-season, and then withdraw the gas in the winter at high rates.

I knew many of the gas producers in the Appalachian Basin and my combined knowledge of technology, law, finance, business and construction allowed me to do much of the project development work myself. I was able to begin injection into the Honeoye Storage facility, in 1975. The facility is still in operation today and I am still its president. Over the years I successfully built another New York facility and laid the ground work for a third. In 2007 I sold my interests in all three of these facilities. I had achieved the American dream after 32 years of operations with these facilities.

The Publishing Company

Dave, let us go back to IHRDC. Tell me more about the beginnings.

In 1972 we opened an office in Boston's Back Bay area. We started teaching courses in computer programming and petroleum technology. As the oil price rose in the 1970s, we had 25 well known industry instructors in all areas of upstream technology. While we were developing the instructional business, a good friend said to me: "*Dave: you should begin to develop products. Lenders do not like to give loans to companies who only offer services*". We started the IHRDC Press. From 1974 to 1978 we published 79 books in upstream technology. It was hard to make a profit because so many invoices remained unpaid. It taught us to invest in new learning products and prepared us for our next major publishing endeavor.

In 1977 there was a major shortage of natural gas in the US and the local gas distribution companies, asked us to help them find and produce gas resources. With two key instructors, we formed the Arlington Group, and participated in an exploration program. We had successfully invested in a portfolio of about 100 prospects over four years, but by 1985, the price dropped sharply and the program was terminated.

In the Meantime: Extension of the Publishing Company

In 1976 video technology evolved and all the major companies built video production studios. We were invited by Phillips Petroleum to record five of our courses in geoscience at their studio in Bartlesville, with what was state-of-the-art video equipment. However, we soon realized that the "bobbing head" presentations were effective only to those who had the patience to look at hour after hour of content.

In 1979 Mobil asked us to develop "Career Development Guides" for their newly hired E&P specialists. These Guides, three for geoscientists and five for petroleum engineers, were the forerunners of today's competency models. They included the list of learning needs of each job linked to the books, Mobil courses or on-job-assignments.

We presented the Guides to Mobil's VP of HR. He asked if I had any recommendations, and I gave him my answer. "*There are no self-study guides available to meet their learning needs. If they miss a Mobil course they have to wait another year. We need to build 145 learning modules consisting of a one-hour mentor videotape and an accompanying manual.*" He answered "We will pay for a pilot program of three modules and see how our specialists respond." I said "No, we will do it on a joint basis sharing the cost 50/50. IHRDC will hold copyright, and you will have the right to reproduce them for your employees." He responded "*Go ahead and produce them on a joint venture basis and, if they are as good as you say they will be, we will buy your company later.*"

Dave and I laugh. The conversation was an example to demonstrate the entrepreneurial spirit of a business transaction at high levels in the oil and gas industry. Then, Dave continued the narrative…

We developed three modules and piloted them with extraordinary good feedback.

The Most Productive Six Months

It took months to convince 10 companies to sponsor the development of the proposed *Video Library for E&P Specialists*. I was able to convince Exxon, Shell, Mobil, Texaco, BP, ENI (AGIP), Aramco, ADNOC, Schlumberger and Phillips Petroleum to commit to sponsorship. We planned the work in seven production cycles of 21 modules each. Each company would pay $100,000 per production cycle and IHRDC would assume the risk of cost overruns. The companies would have the right to use and reproduce the materials for their own employees and IHRDC could sell to the rest of the world.

We began work immediately with three production teams. The sponsors reviewed the video and manual content in advance of production. We had teams of graphic artists, video producers and writers who worked to complete 115 modules in 13 years from 1981 to 1994. It was a massive undertaking.

Happier and Better Performers

We were the biggest producers of instructional videos in the Boston area. Then we saw the evolution of digitization in the early 1990s. The technology of scanning, cataloging and delivering digital content kicked-in, and in 1995, we digitized and cataloged all of the videos and manuals of our *Video Library of E&P* Specialists for easy access on CD-ROMs. When online streaming became possible, we began delivering the content worldwide. Thousands of industry learners in over 64 countries have used this extensive learning system to build their E&P knowledge and skills. I am still in awe of all that we have accomplished.

Two More Major Innovations

What new developments are you advancing now?

We introduced two more significant innovations in recent years: the development of simulation games and the building of competency management systems. Our programs are built around Adult Learning Principles. Our oil, gas and power industry programs have built-in challenging simulation games, in a competency and learning system widely recognized and used by the industry. SPE has adopted 44 of our competency models for its members, and the SEG has adopted 11 of them for its members.

This is a big step forward to support the self-learning of thousands of individuals in oil and gas: As of today, SEG the Society of Exploration Geophysicists with more than 37,000 members and the SPE, the Society of Petroleum Engineers, with about 170,000, offer free to their members a competency assessment, the product of a donation of IHRDC to these organizations.

Of all your achievements and challenges, which were the most important for you?

After I built my gas storage facility in 1975, I built my second in 1991, and began the third in 2007. In that year the price of gas was attractive and companies come to me, offering to buy my gas storage fields facilities at high prices.

An Offer You Cannot Refuse

A new business opportunity for you. But Arlington, the gas storage business, was your baby. How and why did you decide to sell it? Tell me more about this decision.

The first company came to me with an offer for an amount of money that was insufficient, so I was not compelled to sell. A second company offered a little higher price but I also declined. A third company approached me, offering a much larger amount of money, I had to say yes! I had formed a partnership with ownership distributed within my family and with my associate. The American dream really works! What I have learned from my parents and grandparents; that a good education and hard work would pay off, had really happened to me!

The "Wow!" Moment

I had this strong feeling that the advice given to me many years earlier from my grandfather and my parents, at the moment of signing the agreement. I managed to launch the whole project during a low-cost downturn of the gas market, but I had the vision that a day would come, where gas would reach a higher price and storage facilities would be desirable.

My second major accomplishment was to successfully complete the Video Library for E&P Specialists. Definitively! It was a major undertaking.

What interests you now?

If you ask any university if their students are ready for their work upon graduation, they are convinced their fresh graduates are ready for work. The University faculty think 95% of their graduates are ready to start working, but the companies say only 1% of them are ready! A new development is the one occurring at Colleges for America at Southern New Hampshire University.

They offer the opportunity for people to get their degrees online for $3,000 USD per year, for working people who cannot take the time to attend college full time. This interests me!

Another Innovative Idea

I am convinced that we need to evolve into a situation where the on-boarding programs in the oil and gas industry, are offered in such a way that the employees receive recognition for their efforts. It is a huge amount of work, effort, dedication and management by the organization, the company, the training companies, and of course, by the employees. I think all this is worth a Master degree. So, why we do not engage the universities in the loop, and provide a formal certification to that gigantic effort?

Brilliant! The mobility of the new Millennials will trigger that kind of solutions. They would get some extra motivation to stay. NOCs engage in organizing these kinds of on-boarding, but there is a systemic disinterest of the new employees about this training programs. A degree would add that extra motivation.

I think this option could evolve from the professional societies. If SEG, SPE, maybe other professional organizations could certify the competencies, this corporate effort could be worth a Master Degree, perhaps a Master in Petroleum Management and Technology. It would be a professional degree, not an academic degree. We am now teaching a complete, integrated learning program. I think the Master Degree implementation for new hires will evolve from companies like this one. People would welcome and embrace disruptive innovation. Collaborative pioneer, Clayton Christensen, a Harvard Business School Professor, is steering this efforts and vision.

I think he has a Ted Talk about it.

Yes, he talks about innovation. His peer, Dr. Michelle R. Weise is the Executive Director of Sandbox, at Southern New Hampshire University, and she is fully aware of the impact of disruptive Innovation in education. She and her team are working on very interesting developments in this area. Again, this is my new goal: design on-boarding programs for new hires who receive a Master degree for their efforts.

And about challenges? What is your most important challenge?

We always have challenges. I would say one of our continuing challenge is to find sufficient capital to invest in new opportunities. Usually we find funds from internally generated capital, but the drop in oil price has reduced our available capital substantially.

"I Have Never Given Orders"

Do you have any style of leadership?

I very seldom give orders—it just does not happen, especially if you hire competent and educated people. My leadership style consists of setting organizational goals, setting high standards and then allowing managers to make their own plans and decisions, have their own space to make a few mistakes but, in the end, achieve their own successes. Normally if you hire good people, the journey should be fascinating.

The leader must position him or herself ahead of the times, to anticipate trends and drivers of change. Anticipate trends, so as to be ahead of others. You may be ahead of customer acceptance but they soon catch up.

You have to be a team player when you are young. But almost immediately, craft a vision and go for it. When I give a course, I ask people to raise a hand if they would like to have their own company. Almost everyone raise their hands, so I ask them "why you do not have your own company?" No excuses.

You have influence many people along your career and life. Who do you consider to be one of your own role models?

I planned to build an underground gas storage project, and I met Jacek Makowski, who was very helpful in the early days of the project. Jacek had just completed building a large LNG receiving facility. "*What do you want to do*", he asked me. And I said that I was going to build a gas storage facility and sell service to the East Coast gas distribution companies. *I said "I don't need a partner I just want to meet your potential customers".* He replied "I know all of your potential customers. I will introduce them to you only if I am your partner!"

I went to the major gas producing state in the Northeast reviewing gas fields, and entered into an option agreement to purchase one field. I allowed Jacek to become a partner and he proved to be a valuable guide to the utility world, the regulation process and to obtaining debt and equity financing.

What are your personal strengths that propelled your success?

My broad educational background, strong motivation to succeed, commitment to work hard to achieve a goal, confidence, willingness to take a risk, and a creative mind allows me the vision to see opportunities that have a good chance of success. I am not a good manager or administrator. I tend to set goals and standards for the organization and then allow strong people around me to make their own decisions and to succeed. I am a strong proponent of the application of emotional intelligence in the workplace. I think I can get the best out of people.

What are the challenges the energy industry will face in the future?

We have to be much more prepared to manage the impact of global warming. This means that we need to conserve the energy we consume and use fossil fuels on a prudent basis. New technology may help us achieve those goals more easily but the renewable energy costs are coming down fast so integrating gas-fired power plants with renewable energy sources is an obvious solution.

A second major challenge is providing enough food for a future world population of 9 billion people.

What message would you like to send to the young generation?
They should think about their career goals and personal values early in life. They need achieve as much education as possible while they are young to have the freedom to work in different areas as new opportunities come their way. Not be afraid to do things on your own when they are young, to live in different places, take on different jobs and to take risks before they pick up "cargo", this is, a wife or a husband, homes, cars and children… all of which anchor you to a job and place.

A Shared Selfie

- **Your favorite word:** Innovation.
- **A city:** Boston. Such a multicultural, diverse city!
- **An important person of your preference:** I would have said my mother, but after so many, many years, I must say my wife Pam.
- **A personal happy moment:** I am happy about very simple things. Blooming flowers make me happy, happy young children in their innocence make me happy, time with my family makes me happy.
- **Your favorite food:** Cod Fish. I love it, because it has no fat, is not filling. I was raised in a Catholic family, and every Friday, we would eat this fantastic fresh fish.
- **Your favorite color:** Blue!
- **Who supported you the most?** My wife Pamela. She supported me during law school, the early days of IHRDC, the risks we took in developing the first storage field, giving up her career to raise our children, allowing me to travel to develop IHRDC. She was always there with her full support.

We were wrapping up the interview, when suddenly Dave mentions that he has a niece who is a social worker, who used to work in Florida. I understand that he wants to tell me something. It does not require any further trigger. He continues:

- I asked her: 'who are your patients?' "Old people, living in Florida because it is warmer there", she said. I asked her 'Why do you go to see them?' "Because they are depressed and need help". I could not understand, retired people in beautiful Florida depressed? I continued asking her why they were depressed. "Because they never achieved in life what they thought they should have achieved. And now is too late".

I have thought …Wouldn't that be a terrible, terrible way to end your life? Boy! I would hate having to say when I am older that I failed to live up to my potential!"

Dave is 80 years old. We definitively want to have his stamina and vision at 80.

Dr. David T. Donohue, his wife Pamela and Maria A. Capello at the Honors and Awards reception of the Society of Petroleum Engineers, Dubai, October 2016.

Dr. Nansen G. Saleri

"Leadership is always about heart and knowledge".

A Glimpse

Dr. Nansen Saleri is a familiar presence in the Wall Street Journal, Houston Chronicle, Reuters, Bloomberg and CNBC. His insight and perspective are among the most acute in the energy world. There is no major conference or forum providing the business of the sector that would not like to include his voice as a powerful motivator or to trigger deep reflections in the audience. He was cited among the industry's game changers in the book titled, Groundbreakers – The Story of Oilfield Technology and the People Who Made it Happen, by Mark Mau and Henry Edmundson (2015).

A recognized professional and role model of several generations in the oil sector, Nansen cemented his leadership in key roles in top producer companies. He was manager of reservoir engineering for Chevron. Later, he was Head of Reservoir Management for Saudi Aramco, a company that appointed him as its lead spokesperson in worldwide forums in critical moments at the beginning of this century. The global discussion about peak-oil soured the discussions during those years, creating a pessimistic outlook on the future of fossil energy. Dr. Saleri was among other leading figures in energy who led the advocacy for a positive forecast, sharing his conviction grounded on his deep knowledge of the Saudi reservoirs. This slowly but steadily transformed the prevailing perspectives towards a recovery of the confidence of the international markets in the sustainability of the main oil producers of the world. In September 2007, he co-founded Quantum Reservoir Impact (QRI), an elite upstream technology company of which he is the President and CEO. QRI leads projects and operations in China, Mexico, Kuwait and the US.

The oil sector has evolved substantially. Dr. Saleri is dedicating QRI to envision, implement and lead disruptive uplifts in technology for reservoir management. He emphasizes reservoir analytics, a key element that will ensure the enhancement of production in the world, from conventional or non-conventional oil resources.

Dr. Nansen Saleri is a star, not only in his professional profile, but he is also a champion of community service efforts. He volunteers generously for several causes, both in the energy sector as well as in the artistic community. He is the co-founder of the Kristin Saleri Art Foundation, an initiative dedicated to support youth artistic programs and to promote the knowledge about his mother Kristin Saleri's artistic legacy.

- Ph.D., Chemical Engineering, University of Virginia, 1975.
- M.Sc., Chemical Engineering, University of Virginia, 1972.
- B.Sc., Chemical Engineering, Bosphorus University, 1970.
- Holder of three U.S. patents.
- John Franklin Carl Award of the Society of Petroleum Engineers (SPE), 2006.
- SPE Distinguished Member 2006, and SPE Distinguished Lecturer, 1991/1992.
- Member of the Advisory Board of Cyber Engineering of the Houston Baptist University, and the Advisory Board of Petroleum Engineering of the University of Houston.

A Personal Snapshot

Dr. Nansen Saleri's QRI technology consulting firm is associated with many of the main accomplishments of the North Kuwait Directorate of Kuwait Oil Company, initiated during Hosnia Hashim's tenure as Deputy Managing Director, at the time, of that asset. Thus, it would be impossible for her not to know him well. Their business liaison grew to become a permanent mutual respect of the finest kind.

Although we had liaised with Nansen Saleri throughout the years, it did not make the interview easy. Even if we had anticipated his answers to our questions, we were to discover several unknown aspects of a long and complex journey. Several moments during our interview challenged his composure (and ours!). When difficult and unspoken memories surfaced and triggered some sadness, Nansen quickly dissipated the tension with his experienced and courteous comments.

We remain incredulous at the resilience, tenacity and audacity of Nansen Saleri, and we are willing to visit him again.

Arranging the Interview

Nansen lives in Houston, which is known as the energy capital of the world. He welcomed our request of an interview as one welcomes the visit of an old friend, with happiness and a sense of expectation. He wanted to see what was new at our end and to share his latest dreams with us.

He was expecting to participate in the Kuwait Oil Company's 5th "Sharing Best Practices" Conference. This event has become a traditional networking opportunity for the company, to exchange best practices, mainly in reservoir management practices and workflows. Dr. Nansen Saleri will be delivering an opening speech, and he was looking forward to that. We initiated the meeting with a conversation about his approaching travel to Kuwait, in a few months, for the conference.

It was wonderful to reconnect with our colleague of many years.

Everything Is a Continuum

From the US to Kuwait

Nansen, tell us about your initial career. How did it all start?

First and foremost, I want to say that everything is a continuum. We are what our upbringing made of us, a result of our decisions, our mentors, and our choices.

I am a consequence of how my incredibly amazing parents raised me. They were wonderful teachers, mentors and lighthouses for me.

Tell us about your parents, Nansen, where were they from?

My father, Agop, was an engineer, with a double master degree in mechanical and electrical engineering, from the Istanbul Technical University. He also studied at the Ecole Centrale de Paris, in France. Through time, I realized my father was not only an engineer, but most genuinely a true entrepreneur, in both spirit and action. He was deeply involved in the industrialization of Turkey, setting up power-plant complexes throughout central Turkey, with a heavy schedule, handling several projects at the same time. Mother was an established artist of a great stature.

I was brought up in this amazing environment.

Being an Armenian You Had to Excel in Everything

We are three siblings. I have a twin brother, Alen, and a sister, Rehan. Ethnically, we are Armenian. Being a minority in Turkey, we faced some challenges, because of the historical and legacy issues, which are well known. I gained maturity through my parents and through their colorful, sometimes painful, stories.

I learned how to excel in everything. Being an Armenian, you had to excel in everything.

Years passed and I moved around, both for studies and for work, to such an extent, that today I consider myself to be a citizen of the world as much as a proud American.

My parents provided the platform and guidance for us to grow. They were able to make sure that wherever we were, that we would be adding value to our communities. My mother had a very strong leadership role in the Turkish-Armenian circles, specifically in the art circles.

Art is always a very difficult environment in which to become a known name, a recognized figure.

Yes, and it was remarkable, because she was a minority immersed in Turkey's art movements, and she was a well-accomplished artist. She won several international awards. And she was the best story teller I knew.

My parents would go to the movies, and when they came back, my mother would tell us the storyline of the movie. She would act out the main scenes, and stage the high points of sadness or joy for us three. We were enchanted by these stories! On occasion, when I had the opportunity to see the same movies later at the theaters, the stories were not nearly as amusing, as compelling or as intriguing as I thought they would be, judging by my mother's storytelling! Her stories were far better.

We laugh for a long moment, as we imagined the scene of this lady recreating the movie scenes for her children, augmenting the dramatic or funny moments to suit her audience of loved ones. We notice once more that leaders are amazing story-tellers. Nansen is a fantastic story-teller himself, and now we know why. It is in his blood!

To what kind of art was she dedicated?

Abstract figurative arts. Most importantly, painting and ceramics. She had a distinct Kristin Saleri style.

How did your mother influence your character and your leadership?

I was and I am a great admirer of my mom, in particular of her innovative spirit and her tenacity. From my childhood, I spent hours accompanying her at her studio. I learned from the way she would relate with other artists, or how she would organize an exhibition of her pieces. In particular, I learned the way she expressed herself with the colors, always with a secret or mystical theme, and with such a compelling style.

She would keep her artistic activity quite vibrant, while in the meantime raising three children, and being socially very active. I definitively learned a lot from her as well as from my father.

The Smart Move

Tell us about your formative years, your elementary and high school years.

I welcome your question, because I am convinced those formative experiences shaped what is an important part of me.

We all attended an Armenian Catholic elementary school in Pangalti, an area of Sisli, in Istanbul. The Catholic school was very protective, and the teachers were disciplinarians. We felt completely at ease in this system. It was our comfort zone.

Then, my parents did something smart. They moved us out of this comfort zone, and enrolled us in a private Turkish/American middle school. They wanted us to be exposed to the multi-cultural realities of Istanbul. So, we were enrolled in the Ata College in Istanbul for Middle School. Afterwards, I attended Robert College, a prestigious high school (Nobel Prize winning author Orhan Pamuk is a fellow alumni). This was the time when I first experienced an American perspective. I then studied chemical engineering at Robert College (currently the Bosphorus University).

I want to give credit to my education, in particular to Robert College. There, the emphasis was placed on debate and philosophy, on literature, and on open discussions on every subject of study. I consider that Robert College prepared me to be *a* free thinker. I am someone who can discuss or debate any concept. This helped me a lot later on in my career.

"I Was Very Good in Sciences"

Tell us how the decision was made for you to move out from Turkey, to study in the US. We know you graduated from University of Virginia.

My father had this idea that I had to become an Engineer. He insisted on this, because he wanted his sons to be engineers like him. I have two siblings, a twin brother and a sister.

Oh! A twin brother. Did he also study engineering?

No. My brother was a rebel and refused to study engineering. Instead, he studied medicine, and became a psychiatrist.

So, in contrast with my twin brother, I complied with my father's wishes. Plus, I was good in sciences. To me, it made good sense to become an Engineer. It was my father's dream. He himself had wished to go to America.

The Challenge of a Lifetime

After graduation from Robert College Engineering School, I earned a full scholarship to attend the University of Virginia. In a way, I realized that I have fulfilled my father's dream. It included immigrating to America, working as an engineer here, and starting a successful business. Unfortunately, he passed away in 1981, well before witnessing my accomplishments.

I obtained my Master and Ph.D. degrees in chemical engineering in 4 years from the University of Virginia. This period was definitively the challenge of my life.

We had medical complications with our firstborn child, financially stretched on a scholarship budget to raise two children at a time of no credit cards or family backing, the absolute imperative to graduate in four years and find a job (without a green card), etc. Yet this all worked out for a better future in the end.

Tell us about your wife.

A Vital Encounter in Prince's Island

I met my wife on the Prince's Island, Istanbul, the week during which I graduated from High School at age 18. It was an encounter that was so vital for me! Like so many other decisions I have made, our marriage was an instantaneous certainty. From the moment I met her, I was sure she was going to be my wife. She did not know that at the time we met (she was 16), but I was certain.

In ten days after first meeting her, I proposed.

It was a good decision. I am cerebral, but also intuitive. I had a strong sense of what I wanted, and when I met her, I just knew she was the one.

But this is an interesting story, as upon receiving my proposal, she declined.

What?! She really declined? How did that happen?

She was an only child, and she was very attached to her father. She said: "I like you, but I have to ask my father".

I was absolutely disappointed that she would not accept the proposal immediately. Three or maybe four days later, her father had no objections so long as she thought I was the right person.

My marriage has been one of the biggest and most important decisions of my life. We have had 3 children (Sy, Kristin, and Pauline, each with a uniquely intriguing personality) and 5 grandchildren (Dennis, Rhea, Jacob, Colin, and Conner, each acting a decade older than their biological age). It has been a wonderful family journey which is alive and ongoing.

I can divide my career in two main segments: the corporate life and the entrepreneurial one. For the first part, I worked in Chevron and Aramco, reaching some of the highest and most interesting roles you can attain our industry. The second portion is, of course, my company, QRI.

"It Was Her Decision"

How were you hired by Chevron?

There is an interesting anecdote about it. When I finalized my Ph.D, I received two offers, in two different places.

One offer was to work in upstream research, with Chevron in La Habra, in the state of California. The other was to work downstream for the refinery research of Chevron in the Bay Area, in San Francisco. As I had a Ph.D. in Chemical Engineering, I could do both roles.

How did you decide? It must have been difficult!

Not at all. It was my wife who decided! I had to pick La Habra, as she wanted to be near Los Angeles. She was thrilled about the idea of being near such a big center of art and entertainment.

It was not my decision. It was hers.

Quantum Reservoir Impact

Nansen, tell us about your technology consultancy firm, QRI. How did you launch it and how do you relate this to your leadership?

The launching of QRI happened in 2007. We were at a high moment in the oil industry, and no predictions of the 2008 financial crash were forecasted.

The genesis of QRI was a result of another of those decisive moments that happened in my life in a matter of hours. Actually, this decision took one hour, in this case. I met Simon Hodson, our Chairman and principal investor, on September 29, 2006 in Houston. We were certain that a firm dedicated to innovative new thinking that focused on adding value through better reservoir management was needed. I realized Simon was exactly the kind of person with whom I wanted to do business, and we became partners with a handshake. No contracts, no paperwork were needed; in only one hour we agreed to do business together, and QRI was conceived. The company started its operations on September 1, 2007 in Houston with a skeletal crew of seven.

Simon is still our Chairman, and he aligned investors to pitch 15 million dollars in a two-year business plan that was ambitious and visionary. Then 2008 arrived, and with it, the price of oil went down, crashing many of QRI's original plans, and several investors pulled back. Against all odds, we had a second round of investors in QRI, and we were able to move on. QRI braved the storm, and here we are, almost ten years later, with operations in Asia, the Middle East, the Americas, and growing.

I am very proud of the team we have assembled. We have more than 150 employees of the finest profiles you may find in the oil and gas industry.

But the crisis I experienced with QRI in 2008 was an eye-opener for me. It was the most difficult professional struggle in my career so far. There were other struggles more important for me, of personal nature.

September 29, 2002

You mentioned some personal struggles of difficult nature. Would you mind sharing which were those?

The most difficult moment in my life was when our daughter, Kristin, had a traffic accident in Houston that put her into a coma. It was the 29th of September of 2002. I will never forget that day.

This is why I never separate personal from professional life, because I faced a lot of tough situations with my parents. But the toughest moment was when an unknown person knocked at the door one morning at 8 a.m. She told me, *"you have to come with me, your daughter had a serious accident"*. I woke up my wife to tell her something awful happened to our daughter. That was by far the toughest moment in my life.

Kristin was in a coma for a week. We did not know what to expect. We didn't know if she would recover, in the first place, and if so, in what condition she would recover.

A week later, I entered her room at the hospital, and she suddenly woke up and literally asked me *"what's the scoop, dad, what is going on?"* We had a happy ending. I am so grateful about the good outcome of this ordeal and the infinite prayers and good thoughts sent by so many.

This was in 2002, and somehow it captures the essence of what I am telling you. Today she is a fully functional, socially active mother of two. Life and work merge in interwoven ways, which many times you cannot separate. These moments leave an indelible mark in your memory.

Soccer in a Muddy Pitch

Changing gears, Nansen, now tell us, when did you realize you were a leader?

I would say it was through sports. I played soccer, and it was the means to realize that my leadership is an integral of part of who I am.

I was 12 or 13, we were playing 6-a-side soccer in the mud… and someone had to take the leadership role. I felt confident to tell people what to do in the pitch. In the mud, you have to quickly realize who can do what, and we had no coach. So, I started distributing actions about what to do.

Who won?

We did! It was a muddy victory, for sure. It was one to remember.

But my leadership was forged by several events that required someone to step in, to guide, to advise, to coordinate. For example, we had a family crisis when my father at only 55 years of age had a stroke. We were caught completely off guard.

How did you manage? You were a very young man.

You just rise to the situation. Either you are a leader or not. Every time there is a crisis, everyone assumes his or her natural role. In the light of our family crisis, this is the role I assumed. It just felt natural. No one told me to take over the reins of the situation. I felt I should take charge. This has been a constant throughout my life and my career.

The Fame

Part of the reason I became recognized in ARAMCO was because I had strong opinions grounded on my knowledge of the reservoirs' performance and my track record of producing results. In early 2000 onward there was a trend in the media in favor of "Peak Oil" ideas and that the fossil energy times were over. I stood firmly against those arguments based on technical analysis.

My viewpoint was derived from a solid knowledge of the reservoirs. I knew the professional skills of the people in ARAMCO, and it was the utilization of this knowledge of all facets involved, that went against the prevailing trend at the time.

The ARAMCO Front-Line Speaker

The leadership of ARAMCO and the Minister HE Ali Naimi felt that the best persons to represent their company in front of the global media were Mahmoud M. Abdul Baqi, the head of Exploration, and I. A high point was in the global summit of the Center of Strategic and International Studies (CSIS), in Washington, February 24, 2004. We presented "*Fifty-Year Crude Oil Supply Scenarios: Saudi Aramco's Perspective*", which I co-authored with Mahmoud Abdul Baqi, signaling the way ahead for the long sustainability of ARAMCO and the Middle East in the energy scenario for decades to come.

The CSIS is the kind of gathering for top leaders to discern the future. It has been more than a decade, and when one goes back to recheck what was said in those meetings, in relation to what has happened, one can see that our forecasts turned out to be more than correct.

Tell us more about how you created that extraordinary vision.

Well, it was a result of my professional activities. For years, I was responsible for Reservoir Management for Saudi Aramco. I had an intimate knowledge of the performance of the Company's fields. Of course, everything is a team effort and the company's leadership and organizational capabilities were extraordinary. I knew our reservoirs were first class, and that the field performances were top-notch.

Nansen, you used a word, "Intimate", in reference to your knowledge of the reservoirs. That you had an intimate knowledge of the Saudi reservoirs. This demonstrates to us that your mastery in the use of words, in the utilization of strong expressions to exemplify or even to propose an idea. We have noticed leaders are quite often exemplary storytellers. Do you agree?

Leadership is certainly related to the use of powerful communication techniques to inspire, to indicate the path forward, to come to terms, to establish liaisons, and to deliver your ideas. When I became a more experienced professional, I started delivering many talks and presentations. I realized I truly enjoyed giving talks.

The Efficient Leader Knows

I propose that for a leader to be efficient, there are two elements that are of foundational importance: To be a very good communicator and that it matters when the communicator has substance.

But in my opinion, leadership is always about heart and knowledge. I always have been a hands-on type of a leader. That is the kind of CEO I am. In all the projects we do, I want to dig deep. I want to analyze what technology is involved, and how we will create value for the clients.

Which have been your greatest accomplishments to date?

I always like to highlight the improvement of production in Kuwait Oil Company's North Directorate as one of our best accomplishments as QRI. We improved the production, reaching 760,000 barrels of oil per day, starting from 590,000 during an 18-month period. It is an outstanding result, leveraging the team work with all parties involved in the asset, under the leadership of Hosnia Hashim and Emad Sultan.

Another main success of which I am proud is to have gathered the brightest minds in the industry to work in QRI, in order to develop some of the most revolutionary diagnostic technologies for reservoir management. We are like the Apple and Google of the oil and gas industry, in utilizing new computational capabilities in the upstream.

Besides these two accomplishments, I would like to also pick the time when I was awarded the John Franklin Carl Award by SPE, and the invention and application of Maximum Reservoir Contact (MRC) wells in Shaybah and other fields that ushered a new era of innovation in upstream.

No Easy Business in Oil and Gas

Where is more difficult to run business? In what region of the world?
There is no such thing as an easy business. If anyone thinks that there is an easy business, is because she or he has not yet attempted it. If you do not embrace the fact that this is going to be very tough, you are not well prepared. You have to put everything into it! You must invest your best effort, your time, your passion, and your full concentration. If you are not ready to do this, then you are not ready to be an entrepreneur.

How did you manage the ups and downs of QRI?
When things did not go well, I always wanted to know why. And when things went well, it was the same!

Whatever you want to describe as the goal, your employees have to imagine and truly endorse. No room for pretensions!

Your Batting Average

You have been exposed to the media many times. Was there any episode in which you experienced a challenge or mishap in communication?
You are right about the risks of my exposure to the media, as I have given hundreds of interviews. And of course nobody is perfect. Sometimes, when I watch myself on TV or on a web-based interview, I do not like what I said. When I did not do well, I do not make excuses. My understanding and aim is to keep improving. What matters is the batting average. You have to always give 100% effort and hope that you hit 80%.

Another thing that happens with media, and this is why you have to be careful, is that sometimes what you say is taken out of context, with an exaggeration that may distort what you wanted to say. No big deal; that is life.

The Land of Opportunity

What anecdote would you like to tell us about your resilience?
First of all, that success does not come for free, or easily. Resilience helps you every step of the way. One anecdote may exemplify this.

The day I arrived in America, I was with my wife. We were newlyweds, and it was just a week after we got married. Marina used to work for Alitalia, and as a wedding present, we were given two tickets on first class to fly to New York, in a 747 (circa August 1970).

Even if I was very confident, I was also very afraid on the inside. We were newlyweds and only had $650, which was our wedding gift from family and friends. We were shy, and we did not know where to stay. So, from JFK Airport, we took a cab to spend the first night in the big city. We were picked up by an Italian–American cab driver. We started talking with him, and he explained: "*This is America, the land of opportunity. You only have to do two things: be brave and work hard, and you will be successful*".

These words resonate in my memory, and I have realized that all people I admire in life, and in work, have these two qualities. They work hard and they are brave. I have not known a person who has succeeded who has not worked very hard, relying only on his or her brilliant mind.

Whom do you admire the most?

There are so many individuals I admire. Two are from Chevron/Aramco. I met them in Dhahran while on assignment; one is Ed Price. He was the Senior Vice-President of Exploration and Production. What I liked about him is that he was fierce and bold. He knew all the wells, reservoirs, with that intimate knowledge to which we were referring earlier. He was a perfectionist. You have to have some degree of fierceness, and definitively you have to be fair. The other person was also from Chevron/Aramco, Stu Holm, an engineer from whom I gained the criticality for reading and learning continually.

I would have guessed that you learned a lot from the recent phase, when you have met so many leaders at the level of Ministers and Presidents.

Among political leaders, I admire Lincoln for his ethical leadership, George Washington for his willingness to part from power, John Kennedy for his vision, and Mahatma Gandhi for his humanity. Nobody in today's political arena makes my list unfortunately. Speaking from my heart and personal circle, I was truly impressed not only by my parents, but so many countless people and friends. Simon, my friend and partner is at the top. Dr. Kevorkian, a family friend, taught me why it is more important to stand up when things go bad. And about role models, my model in life is Leonardo da Vinci, he was an engineer, an artist, an inventor of so many things! Mona Lisa today represents beauty and perfection, things that we all strive for in life.

A Super-Competitive Era

What is your vision of the challenges for the energy industry of the future?

The challenge is that we are in a super competitive era. If anything, competition is going to increase. And competition in the energy industry is all about efficiency and bringing clean energy to the world.

The goal is to provide clean, environmentally compatible calories and BTUs to consumers. So, today's norms of efficiency will not hold anymore. The energy industry is on the way to becoming very much like the IT industry or the Pharmaceutical industry, with no room for mediocrity. Only the energy companies that can emulate the Googles and Pfizers of the world will still be around in the future.

Will QRI be there in the future?

Yes, my whole vision is for QRI to be the instrument for super-efficiency for the industry, with cutting-edge computational analytics, to make real-time decisions the norm.

What message would you like to send to the young generation?

I would repeat to them what the Italian–American cab driver told me on our first trip to the US: be brave, and work hard! And I would tell them that working hard is not enough. You have to reinvent yourself every day, because the world is becoming very competitive.

Fields of knowledge are changing fast. We can research questions in Wikipedia or Google and find what we seek. So, successful future citizens will be those who continuously do things better and faster. If they are going to rely only on their parents' reputation and their diplomas from top-notch institutions, they will not be successful.

How modern are you? Do you have a Twitter or a Snapchat account?

Not yet, but I will. We are a company that handles trade secrets, and any comment online may have commercial consequences. Nevertheless, I will have to jump in the water, soon.

A Shared Selfie

- **Your favorite word:** Impact.
- **A city:** Istanbul and Houston.
- **An important person of your preference:** Leonardo da Vinci.
- **Your favorite food:** Chocolate ice cream with hot tea.
- **Your favorite color:** I am attracted to a harmony of colors, I like a full spectrum of colors. This is a result of an admiration for my mother's painting, which were typically in hues of fall colors.
- **A favorite song:** Beethoven's 9th Symphony.
- **A personal happy moment:** My daughter Kristin waking up and asking me *"What's the scoop, Dad? What is going on?"*

- **Who supported you the most:** undoubtedly, it has been my wife, Marina. She has always been there for me. She supported my career during the up and down times. Unquestionably she is the best and most important supporter in my life.

Post-scriptum

When we asked Nansen about his personal favorite happy moment, he let us know that his daughter, Kristin, woke from her coma on the same date, September 29, that he reached a handshake agreement to found QRI with Simon Hodson. Two great moments, in 2002 and 2006, on the same September 29.

We guess this is serendipity at its best.

Kristin Saleri Children's Art Competition, Istanbul, Turkey (2015)

After the Interviews: By Way of Epilogue

We definitely learned from each of the interviews and conversations we conducted with all the extraordinary leaders we included in this compilation.

The trenches we felt we transited were shared, as we learned by others. We were not alone experiencing great difficulties, but instead discovered that all what is so shiny in the big role models we like to follow and admire is in reality the coverture of a tangled history, generally made of a leadership capacity to detect opportunities, motivate others, and engage in very hard work. But also the resilience that results from, severe difficulties, loneliness, and even discrimination and direct confrontation.

Several of our assumptions on leadership and resilience were confirmed by the interviews. But others were smashed, when we were assuming some journeys were simple or facilitated. None of the journeys towards success of our interviewed leaders was a simple one.

We found incredible stories of opportunity, like the need of a substitute manager for someone who died. Or the chance to launch a subsidiary of Shell in Jordan, given to a young professional geologist, who had no experience in launching companies. We admired the courage of those professionals sent to operational roles offshore in Alaska, the North Sea, Alaska, and the Gulf of Mexico, in completely different cultures than their own, to lead operational teams or learn their own professional applied skills. And that courage of those who accepted roles out of their core discipline, to lead in new realms.

There were moments of incredible resilience, related to us by leaders who led in foreign countries the efforts of their fellow citizens to regain control of their oil industry. And those of leaders that were indignantly discriminated for being women, or who risked their lives, fiercely resisting to be pushed out by brute force for political reasons. We marveled at the strength and integrity of a CEO who was retired from her job, after she refused to resign after intense campaigns of political maneuvers and press pressure.

We experienced the great insight of these leaders, when they gifted us with remarkable comments or answers. We will always remember that the question to "what is your happiest moment?" was once answered with a plain "to be frank with you, at my age, I have seen the happy and the sad moments of life, so I enjoy every moment". And the incredible young spirit of an 80-years old leader, who did not want "to become old regretting not having fulfilled my dreams".

The trivialities of the most important moments of their life shed a light on us about the humbleness of true leadership. Like when we were told that a call to become Minister arrived while jogging; an appointment as CEO of a corporation arrived simply by phone; and a clerk in a reception desk did not recognized a young president of an oil corporation of the Middle East, and joked with him about duplicity of names.

We rejoiced with them, when they shared the pride they felt about launching the biggest event in oil and gas in the world. And their pride about the success of others; like the case of a leader consultant who drove their clients' merge success, but was not in the scenario with them, signing the final documents. Or the pride of a leader to serve large communities, with a dedicated volunteerism at professional societies for the young generations of engineers in the oil industry.

Our admiration for each one of the leaders we included in this compilation grew. We confirmed leadership is about vision, communication and an innate or acquired ability to inspire others. We also realize leadership is in great deal resilience. Resilience to confront adversity and difficulties, and most importantly, to accept defeat and overcome it. Many times, with no other people cheering about your success but your inner family. Your support circle.

We interviewed 18 giants, who gifted us with 18 new facets for our multilayered comprehension of what is resilience and leadership.

Many of our colleagues and liaisons were not included, as the timeframe we set for ourselves was very tight, and because our compilation could not involve our colleagues, as we are active workers in our respective organizations. We have them in our minds and hearts, and they are part of this compilation, in one way or another.

We, also learned from each other a little bit more, and enjoyed immensely the process of writing this book in a collaborative effort. There were moments when we thought we would not ever finish. But we did! And the finalization of our book left us with the wish to keep compiling learnings. Perhaps with different focus or format, but surely with the same wish of sharing what made of us and others better persons.

We are already looking forward to a new collaboration.

About the Authors

Maria Angela Capello An awarded Leader in Energy, expert in Field Development, Reservoir Management, Talent Management and Women Empowerment. Maria has instrumented and launched innovative implementations in Latin America, USA, and the Middle East. The first woman supervisor of seismic acquisition operations in the flatlands and jungles of Venezuela, she progressed her career to become general manager of an oil asset. Afterwards, she worked for Halliburton in regional roles as Subsurface Practice and Operations Manager for Latin America and the northern Gulf region, Middle East. Maria Angela has pioneered key transformational initiatives for the oil and gas sector, in exploration and production companies as well as in the service companies and consulting firms related to the sector.

She currently works as an executive advisor in Kuwait Oil Company KOC, coordinating the standardization of reservoir management best practices across the company. Additionally, she is the Strategic Lead Advisor of the Professional Women Network for the nine companies of the Kuwait Petroleum Corporation holding, advancing women in the oil sector.

Maria is Italian-Venezuelan and has lived in Kuwait for the past 11 years. Maria is a Distinguished Member and Distinguished Lecturer of the Society of Petroleum Engineers SPE, and received the SPE Distinguished Service Award in 2017. She was Vice-President of SEG, the Society of Exploration Geophysicists, is a member of the SEAM Board of Directors, and has been distinguished with the SEG Lifetime Membership. Maria Angela is a Physicist (Universidad Simon Bolivar, Venezuela), and an MS in Geophysics of the Colorado School of Mines (USA).

Hosnia Hashim A pioneer executive of the Kuwait oil sector, with a career shaped by a trail of successes, Hosnia was the first-ever woman appointed as a director of an oil and gas asset producing 700,000 barrels of oil per day in the State of Kuwait, marking her lastingly a pioneer role model for women in the Middle East and the World.

She is the Deputy CEO of the Petrochemical Industries Company of Kuwait (PIC), in charge of Olefins and Aromatics. She is the Chair and Executive Director of the Board of Equate Petrochemical Company. She also chairs TKOC Board, The Kuwait Olefins Company.

She was Vice President of Operations of Kuwait Foreign Petroleum Exploration Company (KUFPEC), heading operations in 15 countries. She was Deputy CEO of the North Kuwait and West Kuwait assets of Kuwait Oil Company (KOC).

Hosnia is the Founder and Chair of the Professional Women Network of Kuwait Petroleum Corporation and its subsidiaries, a breakthrough initiative for the advancement of women, the first of its kind in the Arabic Gulf.

Hosnia has been featured in Forbes Middle East magazine as one of the most powerful women in Oil and Gas, and was awarded as "Woman of the Year in Oil and Gas" in 2015 by ADIPEC, the Abu Dhabi International Petroleum Exhibition and Conference. A Kuwaiti national, Hosnia chaired the SPE Middle East Board of Directors and was the SPE Regional Director for MENA and India.

Printed by Printforce, the Netherlands